Power Electronic Converters

Interactive Modelling Using Simulink

Power Electronic Converters
Interactive Modelling Using Simulink

By
Narayanaswamy P. R. Iyer

CRC Press is an imprint of the
Taylor & Francis Group, an **informa** business

MATLAB® and Simulink® are trademarks of The MathWorks, Inc. and are used with permission. The MathWorks does not warrant the accuracy of the text or exercises in this book. This book's use or discussion of MATLAB® and Simulink® software or related products does not constitute endorsement or sponsorship by The MathWorks of a particular pedagogical approach or particular use of the MATLAB® and Simulink® software.

CRC Press
Taylor & Francis Group
6000 Broken Sound Parkway NW, Suite 300
Boca Raton, FL 33487-2742

© 2018 by Taylor & Francis Group, LLC
CRC Press is an imprint of Taylor & Francis Group, an Informa business

No claim to original U.S. Government works
Printed on acid-free paper

International Standard Book Number-13: 978-0-8153-6819-9 (Hardback)

This book contains information obtained from authentic and highly regarded sources. Reasonable efforts have been made to publish reliable data and information, but the author and publisher cannot assume responsibility for the validity of all materials or the consequences of their use. The authors and publishers have attempted to trace the copyright holders of all material reproduced in this publication and apologize to copyright holders if permission to publish in this form has not been obtained. If any copyright material has not been acknowledged please write and let us know so we may rectify in any future reprint.

Except as permitted under U.S. Copyright Law, no part of this book may be reprinted, reproduced, transmitted, or utilized in any form by any electronic, mechanical, or other means, now known or hereafter invented, including photocopying, microfilming, and recording, or in any information storage or retrieval system, without written permission from the publishers.

For permission to photocopy or use material electronically from this work, please access www.copyright.com (http://www.copyright.com/) or contact the Copyright Clearance Center, Inc. (CCC), 222 Rosewood Drive, Danvers, MA 01923, 978-750-8400. CCC is a not-for-profit organization that provides licenses and registration for a variety of users. For organizations that have been granted a photocopy license by the CCC, a separate system of payment has been arranged.

Trademark Notice: Product or corporate names may be trademarks or registered trademarks, and are used only for identification and explanation without intent to infringe.

Visit the Taylor & Francis Web site at
http://www.taylorandfrancis.com

and the CRC Press Web site at
http://www.crcpress.com

Dedicated to the memory of the late Dr. Venkat Ramaswamy,

who was a great source of inspiration.

Contents

Preface... xiii

1 Introduction.. 1
   1.1   Background... 1
   1.2   Why Use Simulink? 2
   1.3   Significance of Modelling 2
   1.4   Book Novelty 3
   1.5   Book Outline... 4
   References 5

2 Fundamentals of Interactive Modelling 7
   2.1   Introduction.. 7
   2.2   Interactive Modelling Concept.. 7
   2.3   Interactive Modelling Procedure 8
       2.3.1   Interactive Model Development 9
       2.3.2   Three-Phase AC Voltage Source 12
       2.3.3   SCR Three-Phase Half-Wave Converter 16
       2.3.4   Gate Pulse Generator.. 18
       2.3.5   RLE Load... 21
       2.3.6   Output Voltage and Current Measurement 21
       2.3.7   Input Voltage Measurement 24
   2.4   Simulation Results 27
   2.5   Discussion of Results... 28
   2.6   Conclusions.. 28
   References 29

3 Interactive Models for AC to DC Converters.................................... 31
   3.1   Introduction.. 31
   3.2   Single-Phase Full-Wave Diode Bridge Rectifier............... 31
       3.2.1   Interactive Model for Single-Phase FWDBR with
              Purely Resistive or with RLE Load 32
       3.2.2   Simulation Results 36
   3.3   Single-Phase Full-Wave SCR Bridge Rectifier 38
       3.3.1   Model for Single-Phase FWCBR with Purely
              Resistive or with RLE Load 39
       3.3.2   Simulation Results 44
   3.4   Three-Phase Full-Wave Diode Bridge Rectifier 46
       3.4.1   Model for Three-Phase FWDBR with Purely
              Resistive Load.. 48
       3.4.2   Simulation Results 54

vii

viii   *Contents*

   3.5   Conclusions..56
References 56

4 Interactive Models for DC to AC Converters..57
   4.1   Introduction 57
   4.2   Three-Phase 180° Mode Inverter...57
       4.2.1   Analysis of Line-to-Line Voltage 58
       4.2.2   Analysis of Line-to-Neutral Voltage 60
       4.2.3   Total Harmonic Distortion 63
       4.2.4   Model for Three-Phase 180° Mode Inverter....................63
       4.2.5   Simulation Results 67
   4.3   Three-Phase 120° Mode Inverter 68
       4.3.1   Analysis of Line-to-Line Voltage 71
       4.3.2   Analysis of Line-to-Neutral Voltage 73
       4.3.3   Total Harmonic Distortion 75
       4.3.4   Model of Three-Phase 120° Mode Inverter 76
       4.3.5   Simulation Results 78
   4.4   Three-Phase Sine PWM Technique 80
       4.4.1   Model for Three-Phase Sine PWM Inverter....................84
       4.4.2   Simulation Results 86
   4.5   Conclusions..87
References 90

5 Interactive Models for DC to DC Converters 91
   5.1   Introduction 91
   5.2   Buck Converter Analysis in Continuous Conduction Mode........91
   5.3   Buck Converter Analysis in Discontinuous Conduction Mode.....94
   5.4   Model of Buck Converter in CCM and DCM...............................95
       5.4.1   Simulation Results 97
   5.5   Boost Converter Analysis in Continuous Conduction Mode 101
   5.6   Boost Converter Analysis in Discontinuous Conduction Mode 103
   5.7   Model of Boost Converter in CCM and DCM............................. 105
       5.7.1   Simulation Results 106
   5.8   Buck–Boost Converter Analysis in Continuous Conduction Mode 106
   5.9   Buck–Boost Converter Analysis in the Discontinuous
       Conduction Mode... 112
   5.10 Model of Buck–Boost Converter in CCM and DCM.................... 114
       5.10.1 Simulation Results 115
   5.11 Conclusions... 115
References 120

6 Interactive Models for AC to AC Converters 121
   6.1   Introduction 121
   6.2   Analysis of a Fully Controlled Three-Phase Three-Wire AC
       Voltage Controller with Star-Connected Resistive Load and
       Isolated Neutral 121

Contents ix

6.2.1 Modelling of a Fully Controlled Three-Phase Three-Wire AC Voltage Controller with Star-Connected Resistive Load and Isolated Neutral................124

6.2.2 Simulation Results 129

6.3 Analysis of a Fully Controlled Three-Phase AC Voltage Controller in Series with Resistive Load Connected in Delta 133

6.3.1 Modelling of a Fully Controlled Three-Phase AC Voltage Controller in Series with Resistive Load Connected in Delta 135

6.3.2 Simulation Results 140

6.4 Conclusions................144

References 145

7 Interactive Modelling of a Switched Mode Power Supply Using Buck Converter................147

7.1 Introduction 147

7.2 Principle of Operation of Switched Mode Power Supply............147

7.3 Modelling of the Switched Mode Power Supply 150

7.3.1 Simulation Results 153

7.4 Conclusions................155

References 163

8 Interactive Models for Fourth-Order DC to DC Converters................165

8.1 Introduction 165

8.2 Analysis of SEPIC Converter in CCM................165

8.3 Analysis of SEPIC Converter in DCM................168

8.4 Model of SEPIC Converter in CCM and DCM................171

8.4.1 Simulation Results 174

8.5 Analysis of Quadratic Boost Converter in the CCM................176

8.6 Analysis of Quadratic Boost Converter in the DCM 181

8.7 Model of Quadratic Boost Converter in CCM and DCM............184

8.7.1 Simulation Results 191

8.8 Analysis of Ultra-Lift Luo Converter in the CCM................191

8.9 Analysis of Ultra-Lift Luo Converter in DCM................194

8.10 Model of Ultra-Lift Luo Converter in CCM and DCM................197

8.10.1 Simulation Results 200

8.11 Conclusions................202

References 205

9 Interactive Models for Three-Phase Multilevel Inverters................207

9.1 Introduction 207

9.2 Three-Phase Diode-Clamped Three-Level Inverter 208

9.2.1 Modelling of Three-Phase Diode-Clamped Three-Level Inverter................209

9.2.2 Simulation Results 214

9.3		Three-Phase Flying-Capacitor Three-Level Inverter	214
	9.3.1	Modelling of Three-Phase Flying-Capacitor Three-Level Inverter	218
	9.3.2	Simulation Results	221
9.4		Three-Phase Cascaded H-Bridge Inverter	221
	9.4.1	Modelling of Three-Phase Five-Level Cascaded H-Bridge Inverter	225
	9.4.2	Simulation Results	232
9.5		RMS Value and Harmonic Analysis of the Line-to-Line Voltage of Three-Phase DCTLI and FCTLI	235
9.6		RMS Value and THD of Phase-to-Ground Voltage of TPFLCHB Inverter	237
9.7		Pulse Width Modulation Methods for Multilevel Converters	238
	9.7.1	Multi-Carrier Sine Phase-Shift PWM	239
	9.7.2	Simulation Results	243
	9.7.3	Multi-Carrier Sine Level Shift PWM	243
	9.7.4	Simulation Results	249
9.8		Conclusions	249
References			251

10 Interactive Model Verification 253

10.1		Introduction	253
10.2		AC to DC Converters	253
	10.2.1	Single-Phase Full-Wave Diode Bridge Rectifier	253
	10.2.2	Single-Phase Full-Wave SCR Bridge	254
	10.2.3	Three-Phase Full-Wave Diode Bridge Rectifier	254
10.3		DC to AC Converters	256
	10.3.1	Three-Phase 180° Mode Inverter	258
	10.3.2	Three-Phase 120° Mode Inverter	258
10.4		DC to DC Converter	260
	10.4.1	Buck Converter	260
	10.4.2	Boost Converter	262
	10.4.3	Buck–Boost Converter	265
10.5		AC to AC Converter	265
	10.5.1	Three-Phase Thyristor AC to AC Controller Connected to Resistive Load in Star	267
	10.5.2	Three-Phase Thyristor AC to AC Controller in Series with Resistive Load in Delta	270
10.6		Switched Mode Power Supply Using Buck Converter	272
10.7		Fourth-Order DC to DC Converters	280
	10.7.1	SEPIC Converter	282
	10.7.2	Quadratic Boost Converter	283
	10.7.3	Ultra-Lift Luo Converter	283

Contents xi

10.8 Three-Phase Three-Level Inverters 290
     10.8.1 Three-Phase Diode-Clamped Three-Level Inverter 292
     10.8.2 Three-Phase Flying-Capacitor Three-Level Inverter 292
10.9 Three-Phase Sine PWM Inverter 297
10.10 Three-Phase Five-Level Cascaded H-Bridge Inverter 300
10.11 Pulse Width Modulation Methods for Multilevel Converters 310
     10.11.1 Multi-Carrier Sine Phase-Shift PWM 310
     10.11.2 Multi-Carrier Sine Level Shift PWM 315
10.12 Conclusions 318
References 319

**11 Interactive Model for and Real-Time Simulation of a
Single-Phase Half H-Bridge Sine PWM Inverter** 321
11.1 Introduction 321
11.2 Interactive Model of Single-Phase Half H-Bridge Sine
PWM Inverter 322
     11.2.1 Simulation Results 322
11.3 Real-Time Software in the Loop Simulation 322
     11.3.1 Digital Signal Processor 322
     11.3.2 Code Composer Studio 324
     11.3.3 Symmetric PWM Waveform Generation 324
     11.3.4 Sine-Triangle Carrier PWM Generation 328
11.4 Conclusions 332
References 333

Index 335

Preface

This textbook provides a step-by-step method for the development of a virtual interactive power electronics laboratory. Therefore, this textbook is suitable for undergraduates and graduates in electrical and electronics engineering for their laboratory courses and projects in power electronics. This textbook is equally suitable for practising professional engineers in the power electronics industry. They can develop an interactive virtual power electronics laboratory as outlined in this textbook and perform simulation of their new power electronic converter designs. Similarly, instructors may find this textbook useful for running an interactive virtual laboratory course or a new project in the area of power electronics.

The topic of power electronic converter modelling is a growing field. There are many software packages available for this purpose. Most of them are power electronic circuit simulation packages using semiconductor and other passive components. Circuit component-level simulation of power electronic converters is a basic or fundamental level of modelling. A higher level of modelling is the system model, which solves the characteristic equations describing the behaviour of the power electronic converter. The Simulink® software, developed by The Mathworks Inc., USA, can be used to develop advanced system models as well as circuit component-level models for any given power electronic converter. Moreover, it is possible to develop both interactive system models and circuit component-level models using Simulink, which is a special feature. Therefore, I have used Simulink to develop interactive models of power electronic converters presented in this textbook. Unless specified otherwise, the term *model* in this textbook refers to Simulink model.

Chapter 1 provides the introduction, where the significance of modelling power electronic converters, the novelty of this textbook and the reason for the choice of the Simulink software are presented. The fundamentals of developing interactive models are presented in Chapter 2 with an example of a three-phase half-wave thyristor rectifier converter. This chapter also highlights the advantage of interactive modelling, which is primarily the saving in time in testing power electronic converters by model simulation. Interactive system models of power electronic converters are presented in the following order:

AC to DC, DC to AC, DC to DC and AC to AC converters are presented in Chapters 3 to 6, respectively. The buck converter switched mode power supply is presented in Chapter 7. This is followed by selected fourth-order DC to DC converters in Chapter 8 and, finally, three-phase multilevel inverters in Chapter 9.

xiii

To verify the performance of the system models of the power electronic converters in Chapters 3 to 9, Chapter 10 has been added. Chapter 10 provides the interactive component-level models of the power electronic converters presented in Chapters 3 to 9.

An interactive component-level model and real-time software in the loop (SIL), also known as processor in the loop (PIL) simulation of a single-phase half H-bridge sine pulse width modulation (PWM) inverter is presented in Chapter 11. This work was carried out by the author while a research fellow in the Faculty of Engineering, University of Nottingham, UK. The author is grateful to Prof. Alberto Castellazzi and Prof. Pat Wheeler, both of the School of Electrical and Electronics Engineering, University of Nottingham, UK, for their support in carrying out this work.

As mentioned before, it is possible to develop system models of power electronic converters using the Simulink block set and circuit component-level models using the Simscape-Electrical, Power Systems and Specialized Technology block sets.

Much of the material presented in this textbook is work undertaken by the author during his Master of Engineering degree by research programme at the University of Technology Sydney, NSW, Australia. The author is grateful to the late Dr. Venkat Ramaswamy, formerly senior lecturer, School of Electrical Engineering, University of Technology Sydney, NSW, who was a great source of encouragement and support. The author also wishes to thank Dr. Jianguo Zhu, head, School of Electrical Engineering, University of Technology Sydney, NSW, for his support.

The whole of Chapter 8 and the models in Sections 5.4, 5.7 and 5.10 and in Sections 9.4 and 9.7 are my original contributions and are not a part of the above research work.

No end of chapter exercises are given, as this seems to be unnecessary. The reader can practice interactive modelling according to the method outlined here in this book and using the many textbooks on power electronics, some of which are mentioned in the references at the end of each chapter.

The step-by-step interactive modelling details are explained in Chapter 2 only. They are not repeated in Chapters 3 to 11, as most of the sources and components used in the power electronic converter shown in Chapter 2 are repeatedly used in the converters presented in these chapters.

The whole of Chapter 10, excluding Sections 10.7, 10.8, 10.10 and 10.11, is suitable for a senior-level undergraduate curriculum in power electronics.

The model files in this textbook can be found on the website https://www.crcpress.com/Power-Electronic-Converters-Interactive-Modelling-Using-Simulink/Iyer/p/book/9780815368199. Instructors can download a copy of these model files.

I presume that students, instructors and practising professional engineers in the industry alike will find this textbook useful.

Preface

xv

I would like to thank Ms. Nora Konopka, Editorial Director: Engineering, CRC Press for her help and support. Also, many thanks to the anonymous reviewers for their valuable comments.

Finally, I would like to thank my wife, Mythili Iyer, for her patience and understanding. Without her support, this work would not have been possible.

Narayanaswamy P. R. Iyer
Sydney, NSW
Australia

MATLAB® is a registered trademark of The MathWorks, Inc. For product information, please contact:

The MathWorks, Inc.
3 Apple Hill Drive
Natick, MA 01760-2098 USA
Tel: 508 647 7000
Fax: 508-647-7001
E-mail: info@mathworks.com
Web: www.mathworks.com

1

Introduction

1.1 Background

The performance of power electronic converters can be studied using mathematical equations which describe the behaviour of the particular converter. For example, the load current and load voltage pattern for a single-phase full-wave diode bridge rectifier delivering power to a series-connected R–L load can be studied by developing and then solving the differential equation describing the behaviour of this converter. The power electronic converter supplying power to these loads can be developed either as a system model that duplicates the performance of this converter or as a circuit model using the actual power electronic semiconductor and other passive components.

The models developed in this book are system models that solve the equations describing the behaviour of the system. A relatively new concept known as the *switching function concept* is used to simulate the behaviour of the power electronic converters [1–9]. At this stage, it will be appropriate to mention the difference between the system model and the circuit component-level model that is obtained by electronic circuit simulation software. While the system model solves the characteristic equation describing the behaviour of the system, circuit component-level models developed using electronic circuit simulators use the passive and active semiconductor component parameters, such as, for example, ON and OFF switch resistance, voltage and current gain values, parasitic or stray capacitances between junctions, junction potentials, inherent inductive reactance and so on arranged as a subcircuit (also called the *equivalent circuit* of the semiconductor component or the device) to solve a given power electronic converter. In the system models, the characteristic equation describing the behaviour of the system can be either algebraic or differential.

1.2 Why Use Simulink?

In this book, I have used Simulink® software developed by The Mathworks Inc. [10]. Simulink has the following advantages compared with other software packages with regard to power electronic circuit simulation:

- Facility to develop interactive models with user dialogue boxes for power electronic systems, with which any given power electronic converter can be tested by entering the parameters in the dialogue boxes without actually going into each block of the model to enter the data. This saves time for the user and finds applications in the virtual power electronics laboratory. This is one of the unique features in Simulink.
- Harmonic analysis of power electronic converters can be easily developed with Powergui in the PowerSystems block set.
- Many non-linear phenomena such as pulse width modulation (PWM) techniques can be easily verified.
- Many power electronic circuits can be studied by modelling techniques and can be verified by semiconductor components in the PowerSystems block set; that is, it is possible to develop both system models based on the characteristic equations describing the behaviour of the system and electronic circuit component-level models utilising the equivalent circuit parameters of the semiconductor component.
- Facility to study analogue and digital gate drives used in power electronic circuits.
- Facility to save data in workspace, which can be later brought to the MATLAB® window for further mathematical processing, editing graphics and so on.

1.3 Significance of Modelling

The significance of the term *modelling* is summarized here:

- Modelling a physical system refers to the analysis and synthesis to arrive at a suitable mathematical description encompassing the dynamic characteristics of the system in terms of its parameters [11].
- Models are used to predict the performance of the given system [11].

Introduction

- Model prediction permits engineers to think of its potential applications and practical implementation and to develop various control strategies [11].
- Reduces time involved or shortens the overall design process [12, 13].
- Saves time and money as compared with procuring, installing and testing the system in the laboratory, especially when the system is too bulky [11].
- Simulation refers to performing experiments on the model [11].
- Computer simulation plays a vital role in the R&D of power electronic devices for its high manoeuvrability, low cost and ability to speed up system implementation [14].

1.4 Book Novelty

The book mainly concentrates on interactive modelling of selected power electronic converters using Simulink. This book's novelties are given here:

1. The interactive system model for single-phase AC to DC converter using diodes developed here is new and different from that existing in the literature references.
2. The interactive system model for single-phase AC to DC converter using silicon-controlled rectifiers (SCRs) developed here is new.
3. The interactive system model for three-phase inverter in the discontinuous current conduction (120°) mode is new.
4. The interactive system model three-phase sine-triangle carrier PWM inverter is new.
5. The interactive system models for three-phase thyristor AC to AC controllers are new.
6. The interactive system model for buck converter switched mode power supply (SMPS), especially the feedback control part using a triangle carrier is new. This is different from that existing in the literature references, where a saw-tooth carrier is used.
7. The interactive system model for quadratic boost and ultra-lift Luo converters are new.
8. The interactive system models developed for three-phase diode-clamped three-level inverter (DCTLI), flying-capacitor three-level inverter (FCTLI) and five-level cascaded H-bridge inverter (FLCHBI) are new.

4   *Power Electronic Converters*

9. The interactive system models developed for multi-carrier sine-phase shift PWM (MSPSPWM) and multi-carrier sine-level shift PWM (MSLSPWM) are new.

10. The software in the loop (SIL) or processor in the loop (PIL) simulation is presented with a power electronic converter example.

11. System models of power electronic converters are verified for performance using interactive circuit component-level models developed using Simscape-Electrical, Power Systems and Specialized Technology block sets.

1.5 Book Outline

In this text book, interactive system models for power electronic converters have mainly been developed and are then verified using interactive circuit component-level models.

Chapter 1 provides the introduction. In Chapter 2, the method of developing and advantage of interactive modelling are presented with a power electronic converter example. Chapters 3 through 6 provide building of interactive system models for AC to DC, DC to AC, DC to DC and AC to AC converters respectively. In Chapter 3, system models for single-phase full-wave diode bridge rectifier (FWDBR), single-phase full-wave thyristor silicon controlled rectifier (SCR), controlled bridge rectifier (FWCBR) and three-phase FWDBR are presented. In Chapter 4, system models for continuous current conduction (180°) mode and discontinuous current conduction (120°) mode inverters and the three-phase sine-triangle carrier PWM inverter are presented. In Chapter 5, system models for second-order DC to DC converters such as buck, boost and buck–boost converters are presented. In Chapter 6, three-phase thyristor AC to AC controllers connected in series with resistive loads in star with isolated neutral and three-phase thyristor AC to AC controllers in series with resistive load in delta are presented. In Chapter 7, the system model for SMPS using a buck converter is presented. The system models for fourth-order DC to DC converters such as single-ended primary inductance converters (SEPIC), quadratic boost and ultra-lift Luo converters are presented in Chapter 8. Chapter 9 deals with system models for three-phase DCTLI, three-phase FCTLI, three-phase FLCHBI, multi-carrier sine-phase shift (MSPS) and multi-carrier sine-level shift (MSLS) PWM. Finally, system models of power electronic converters discussed in Chapters 3 through 9 are verified for performance using circuit component-level models in Chapter 10, where semiconductor and passive components from Simscape-Electrical, Power Systems and Specialized Technology block

Introduction 5

sets are used. An interactive component-level model and real-time SIL or PIL simulation of a single-phase half H-bridge sine PWM inverter is presented in Chapter 11.

In the system- and component-level models for these power electronic converters, it is the aim to simulate any given power electronic converter by entering the parameters of the converter into the appropriate dialogue boxes, without altering the inner details of the model. This easy-to-use system, and component-level models, save time and are suitable for virtual power electronic laboratory applications.

References

1. B.K. Lee and M. Ehsani: "A simplified functional model for 3-phase voltage source inverter using switching function concept", *IEEE-IECON '99*; Vol.1; San Jose, CA, November–December 1999; pp. 462–467.
2. B.K. Lee and M. Ehsani: "A simplified functional simulation model for three-phase voltage source inverter using switching function concept", *IEEE Transactions on Industrial Electronics*; Vol.48, No.2, April 2001; pp. 309–321.
3. V.F. Pires and J.F.A. Silva: "Teaching nonlinear modelling, simulation and control of electronic power converters using MATLAB/SIMULINK", *IEEE Transactions on Education*; Vol.45, No.3, August 2002; pp. 253–256.
4. B. Baha: "Modelling of resonant switched-mode converters using SIMULINK", *IEE Proceedings, Electric Power Applications*; Vol.145, No.3, May 1998; pp. 159–163.
5. G.D. Marques: "A simple and accurate system simulation of three-phase diode rectifiers", *IEEE-IECON*; Aachen, Germany, 1998; pp. 416–421.
6. B. Baha: "Simulation of switched-mode power electronic circuits", *IEE International Conference on Simulation*, United Kingdom, 1998; pp. 209–214.
7. A.N. Melendez, J.D. Gandoy, C.M. Penalver, and A. Lago: "A new complete nonlinear simulation model of a buck DC–DC converter", *IEEE-ISIE'99*; Slovenia, 1999; pp. 257–261.
8. H.Y. Kanaan and K. Al-Haddad: "Modeling and simulation of DC-DC power converters in CCM and DCM using the switching functions approach: Applications to the buck and Cuk converters", *IEEE-PEDS 2005*; Malaysia, November–December 2005; pp. 468–473.
9. H.Y. Kanaan, K. Al-Haddad, and F. Fnaiech: "Switching-function-based modeling and control of a SEPIC power factor correction circuit operating in continuous and discontinuous current modes", *IEEE International Conference on Industrial Technology*, Tunisia, 2004; pp. 431–437.
10. The Mathworks Inc.: "MATLAB/Simulink", R2016b, 2016.
11. C.-M. Ong: *Dynamic Simulation of Electric Machinery Using MATLAB/SIMULINK*, Upper Saddle River, NJ: Prentice Hall; 1998.
12. N. Mohan, T.M. Undeland and W.P. Robbins: *Power Electronics: Converters, Applications and Design*, Hoboken, NJ: Wiley; 1995; Chapter 4, pp. 61–76.

13. N. Mohan, W.P. Robbins, T.M. Undeland, R. Nilssen and O. Mo: "Simulation of power electronic and motion control systems", *Proceedings of the IEEE*, Vol.82, No.8, August 1994; pp. 1287–1292.
14. X. Xie, Q. Song, G. Yan, and W. Liu: "MATLAB-based simulation of three level PWM inverter-fed motor speed control system", *IEEE APEC '03*; Florida, February 2003; Vol.2; pp. 1105–1110.

2

Fundamentals of Interactive Modelling

2.1 Introduction

In this chapter, fundamentals relating to the interactive modelling concept are introduced. Interactive modelling is a facility available in the Simulink® software. With an interactive model, the user can simulate and solve the problem in hand without going into each of the model blocks to enter parameters. If the same or different users want to simulate the very same power electronic converter topology with different parameters, then it is easy to use the interactive model. The same user or different users can enter their data in the appropriate dialogue boxes and solve their power electronics design problems. This saves time, as the user does not have to click each module in the model to enter the data. The method of interactive model development for power electronic converters is illustrated with an example.

2.2 Interactive Modelling Concept

Interactive modelling is a facility available in the Simulink software [1]. Interactive modelling is suitable for use in academic research and virtual power electronics laboratories in universities, as well as in a power electronic converter design and manufacturing environment. Imagine a power electronics industry manufacturing different types of DC to DC converters. Assume that different clients order the same DC to DC buck converter, each specifying different input voltage, output voltage, output power, output voltage and inductor current ripple and switching frequency requirements. With component values designed by the industry for each converter specification using the well-known modelling equations for the buck converter, the industry can easily and quickly test the performance of each of the buck converters using an interactive model. This is because the interactive model provides dialogue boxes where the designer can easily enter parameter values and test their performance in less time as compared with a conventional model where

7

8   *Power Electronic Converters*

the designer has to open each component subsystem in the model to enter their parameter values. With different specifications for the buck converter, the time consumed for testing with a conventional model will be greater compared with testing the same with an interactive model. Thus, interactive modelling accelerates the manufacturing process and saves time and money.

2.3 Interactive Modelling Procedure

The interactive modelling procedure depends on the topology of the power electronic converter such as the value of the DC voltage source, the amplitude, frequency and phase of single-phase or three-phase AC source, pulse amplitude, frequency, pulse width and phase delay for the gate drive used for the semiconductor switches, component values for resistors, inductors and capacitors used in the model. These are tabulated in Table 2.1.

Interactive modelling involves assigning a variable name to each of the parameter values in Table 2.1 that are used in the model. In some cases, parameters such as on–off resistance and junction forward voltage for semiconductor switches may be built into the model already available, and in such cases, these can be neglected. Then, identical sources or components that form one group are selected by clicking the mouse and the subsystem is created from a selection using the 'Diagram' menu. Then, for the subsystem selected, the 'Create Mask' option from the 'Diagram' menu is used to develop the interactive model. 'Create Mask', or 'Edit Mask' if the mask is already created, opens up the Mask Editor. Then, in the Mask Editor, the 'Parameters and Dialog' menu is selected. In this menu, the user clicks the 'Edit' button, then enters the parameter description and parameter group variable name. The parameter group variable name entered must be the one

TABLE 2.1

Interactive Model Parameters

Sl. No.	Source/Component	Parameter Values
1	DC voltage source	Amplitude
2	Single-phase AC voltage source	Amplitude, frequency, phase delay
3	Three-phase AC voltage source	Amplitude, frequency, phase angle, phase delay
4	Gate pulse drive	Amplitude, frequency, pulse width, phase delay
5	Resistor, inductor, capacitor, diode	Values in ohms, henries, farads, on–off resistance and junction forward voltage
6	Semiconductor switch (BJT, IGBT, MOSFET, SCR)	Gate threshold voltage, on–off resistance and junction forward voltage

Fundamentals of Interactive Modelling 9

used for the source/component in the subsystem. This is repeated for each parameter used in the created subsystem. After clicking the 'Apply' button, the user selects the 'Initialization' menu from the Mask Editor. Any value specification or condition to be fulfilled for parameter variables is entered in the 'Initialization commands' box. Finally, the 'Documentation' menu in the Mask Editor is selected, where the name or title of the subsystem and a brief description relating to the numerical value showing the unit to be entered in each of the pop-up menus or dialogue boxes are entered. Then, the 'Apply' and 'OK' buttons are clicked to complete the interactive model development. In the next section, the interactive model development is explained with an example.

2.3.1 Interactive Model Development

In this section, the interactive model development process is explained with an example of a power electronic converter model. This method is the same irrespective of whether a power electronic converter component-level or system-level model is used. To encompass all the sources/components specified in Table 2.1, a three-phase half-wave silicon controlled rectifier (SCR) converter is considered here [2]. Figure 2.1 shows a three-phase SCR half-wave converter. A three-phase 120 V root mean square (RMS) line-to-neutral, 60 Hz AC voltage source is used. The parameters used for the model are tabulated in Table 2.2.

In Figure 2.1, the three-phase AC voltage sources R, Y and B are connected to the anodes of SCRs SCR1, SCR2 and SCR3, respectively. The cathodes of these three SCRs are tied together and connected to a single-phase load comprising R1, L1 and E. The gate pulse firing angle α for SCR1 in Phase R is 90°. The gate pulse firing angle for SCR2 in Phase Y will be $(\alpha + 120)$ and that for SCR3 in Phase B will be $(\alpha + 240)°$. A suitable amplitude for the gate pulse depends on the SCR gate turn-on threshold voltage, and a suitable gate pulse width is provided that depends on the data sheet for the relevant SCR. These gate drives for the three SCRs are not shown in Figure 2.1 for clarity. The interactive model development for this converter is explained next.

Figure 2.2 shows the interactive model of the three-phase SCR half-wave controller and Figure 2.3 shows the same model with interactive blocks. Figure 2.4a–f shows the various subsystems used in the model shown in Figure 2.2. These are explained as follows:

The three-phase AC voltage source subsystem is shown in Figure 2.4a. The three single-phase AC sources marked AC Voltage Source, AC Voltage Source1 and AC Voltage Source2 form the three-phase AC Voltage Source block. The three-phase half-wave SCR converter subsystem is shown in Figure 2.4b, and consists of three SCRs marked Thyristor, Thyristor1 and Thyristor2. The gate pulse generator subsystem is shown in Figure 2.4c, and consists of three gate drives marked Pulse Generator, Pulse Generator1 and Pulse Generator2. The RLE (R–resistor; L–inductor; E–EMF source, where E can be a DC voltage source, battery or AC voltage source and has to

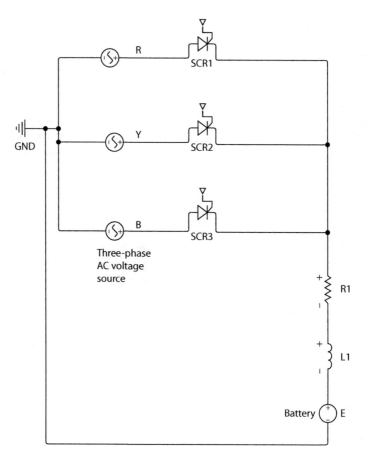

FIGURE 2.1
Three-phase half-wave SCR converter.

TABLE 2.2

Model Parameters

Sl. No.	Parameter	Value	Unit
1	RMS line-to-neutral input voltage	120	Volts
2	Frequency	60	Hertz
3	Gate drive amplitude	10	Volts
4	Gate drive frequency	60	Hertz
5	SCR firing angle for reference phase	90	Degrees
6	Load resistance	0.5	Ohms
7	Load inductance	6.5e−3	Henries
8	Load battery DC voltage	10	Volts

Fundamentals of Interactive Modelling

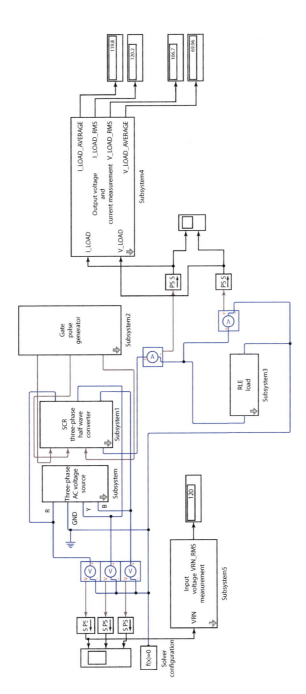

FIGURE 2.2
Three-phase half-wave SCR converter.

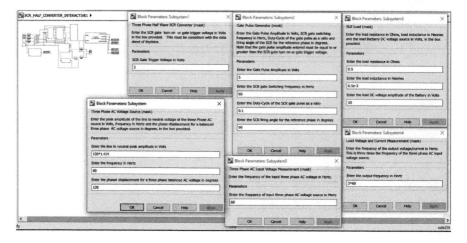

FIGURE 2.3
Three-phase half-wave SCR converter with interactive blocks.

be understood by referring to the circuit schematic) load subsystem is shown in Figure 2.4d, consisting of a resistor, inductor and a DC voltage source all connected in series. The output voltage and current measurement subsystem is shown in Figure 2.4e, where RMS and MEAN value measurement blocks are used to display the RMS and mean or average value of the output voltage and currents. The input AC voltage measurement block is shown in Figure 2.4f, consisting of an RMS measurement block to display the RMS value of the line-to-neutral input voltage. Detailed interactive model formation is discussed in the following sections.

2.3.2 Three-Phase AC Voltage Source

The three-phase AC voltage source subsystem is shown in Figure 2.4a. The dialogue box relating to the AC voltage source module connected to Phase R is shown in Figure 2.5a. The peak amplitude variable is entered as 'A' in volts. The phase shift is entered as 'phi*0' in degrees. The frequency is entered as 'f' in hertz. In the AC Voltage Source1 module relating to Phase Y and the AC Voltage Source2 module relating to Phase B, the entry corresponding to the phase shift is entered as 'phi*(−1)' and 'phi*(1)' (or 'phi*(−2)'), respectively, with all other entries as previously. The following procedure is used to make the three-phase AC voltage source model interactive.

- *Step 1:* In Figure 2.2, before masking, which appears like Figure 2.4a, select AC Voltage Source, AC Voltage Source1 and AC Voltage Source2 using the mouse pointer. Then, click Diagram → Subsystem and Model Reference → Create Subsystem from the selection to form a subsystem called 'Three-Phase AC Voltage Source'.

Fundamentals of Interactive Modelling

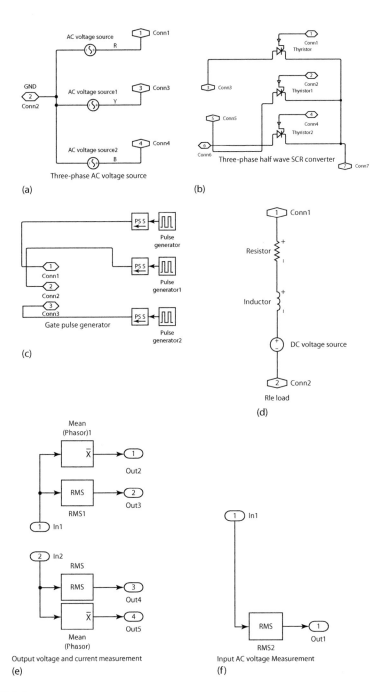

FIGURE 2.4
(a) Three-phase AC voltage source. (b) Three-phase half-wave SCR converter. (c) Gate pulse generator. (d) RLE load. (e) Output voltage and current measurement. (f) Input AC voltage measurement.

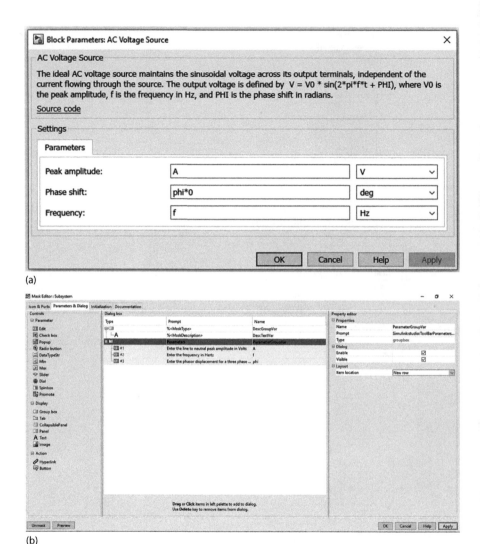

FIGURE 2.5
(a) AC voltage source parameters. (b) 'Parameters and Dialog' menu for three-phase AC voltage source.

- *Step 2:* Select the subsystem marked 'Three-Phase AC Voltage Source', and then click Diagram → Mask → Create Mask. This opens the Mask Editor.
- *Step 3:* In the Mask Editor, click Parameters and Dialog → Edit.
- *Step 4:* Click or select 'Parameters' and enter the instructions to the user.

Fundamentals of Interactive Modelling 15

- *Step 5:* Click or select 'Parameter Group Var' and enter the variable name used in the subsystem for the particular parameter.
- *Step 6:* Repeat Steps 4 and 5 until all parameter variable names are entered for the particular subsystem. Then click the 'Apply' button.

  A view of the 'Parameters and Dialog' menu completed for the 'Three-Phase AC Voltage Source' subsystem is shown in Figure 2.5b.
- *Step 7:* Click 'Initialization' in the Mask Editor.

FIGURE 2.5
(c) 'Intialization' menu for three-phase AC voltage source. (d) 'Documentation' menu for three-phase AC voltage source.

16   *Power Electronic Converters*

- *Step 8:* Ensure all parameter group variables are displayed on the left-hand side.
- *Step 9:* In the 'Initialization commands' box, enter the condition to be satisfied for each parameter variable. This can be a parameter variable greater than zero or lying within a range or above, below or equal to a particular value. Then click the 'Apply' button.
- A view of the 'Initialization' menu completed for the 'Three-Phase AC Voltage Source' subsystem is shown in Figure 2.5c.
- *Step 10:* Click 'Documentation Menu' in the Mask Editor.
- *Step 11:* In the 'Type' box, enter the name of the subsystem.
- *Step 12:* In the 'Description' box, enter the details of the subsystem, the pop-up menu and the value to be entered in each pop-up menu with units.
- *Step 13:* In the 'Help' box, provide any other information relating to the subsystem that the user may need. Clicking the 'Help' button opens up this dialogue box. If there is no other information, the 'Help' box can be left blank. Then click the 'Apply' and 'OK' buttons.

The 'Documentation' menu for the three-phase AC voltage subsystem is shown in Figure 2.5d.

2.3.3 SCR Three-Phase Half-Wave Converter

The three-phase half-wave SCR converter subsystem is shown in Figure 2.4b. The dialogue box relating to Thyristor connected to Phase R is shown in Figure 2.6a. Here, only one parameter, which is the gate trigger voltage, is entered as 'vth' in volts. The thyristor will be turned ON only if the gate cathode voltage exceeds this gate trigger voltage 'vth'. For Thyristor1 and Thyristor2 in phases Y and B, the same value, 'vth', is entered for the gate trigger voltage. The following procedure is used to make the SCR three-phase half-wave converter model interactive.

Step 1: In Figure 2.2, select the SCRs marked Thyristor, Thyristor1 and Thyristor2 using the mouse pointer. Then, click Diagram → Subsystem and Model Reference → Create Subsystem from Selection to form a subsystem called 'SCR Three-Phase Half-Wave Converter'. Then, follow Steps 2 to 6 in Section 2.3.2. A view of the 'Parameters and Dialog' menu completed for the SCR three-phase half-wave converter subsystem is shown in Figure 2.6b.

Then, follow Steps 7 to 9 in Section 2.3.2 to open and complete the 'Initialization commands' box. The 'Initialization commands' box completed for the SCR three-phase half-wave converter is shown in Figure 2.6c.

Finally, follow Steps 10 to 13 in Section 2.3.2 to complete all the dialogue boxes in the 'Documentation' menu. A view of the 'Documentation' menu completed for the SCR three-phase half-wave converter subsystem is shown in Figure 2.6d.

Fundamentals of Interactive Modelling

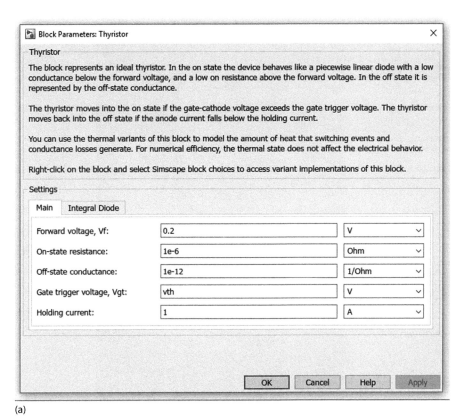
(a)

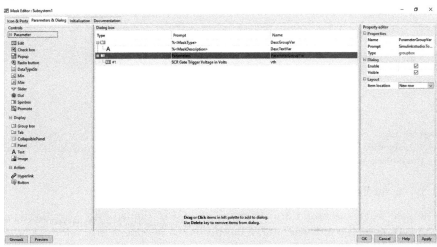
(b)

FIGURE 2.6
(a) Thyristor parameters. (b) 'Parameters and Dialog' menu for three-phase half-wave SCR converter.

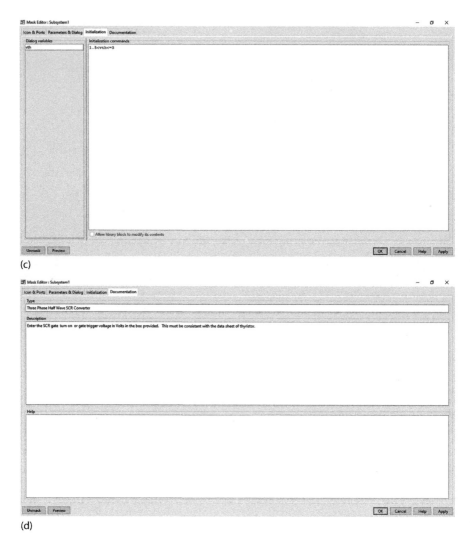

FIGURE 2.6
(c) 'Initialization' menu for three-phase half-wave SCR converter. (d) 'Documentation' menu for three-phase half-wave SCR converter.

2.3.4 Gate Pulse Generator

The gate pulse generator subsystem is shown in Figure 2.4c. The dialogue box relating to the pulse generator connected to the gate of Thyristor in Phase R is shown in Figure 2.7a. Here, the various parameters are pulse amplitude 'VG' in volts, pulse period 1/(f) in seconds where 'f' is the frequency of the gate pulse in hertz, pulse width or duty cycle as a ratio of the pulse period and phase delay alfa in degrees. In the dialogue box in Figure 2.7a, 'Amplitude' is entered as 'VG' V, 'Period' is entered as 1/(f) s, 'Pulse Width' is

Fundamentals of Interactive Modelling

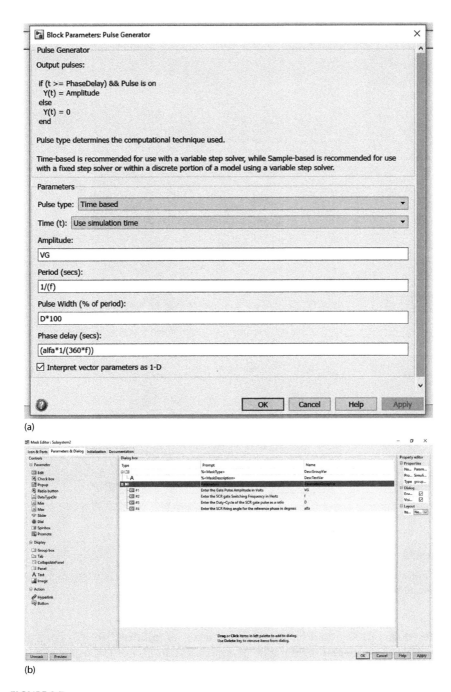

FIGURE 2.7

(a) Pulse generator parameters. (b) 'Parameters and Dialog' menu for gate pulse generator.

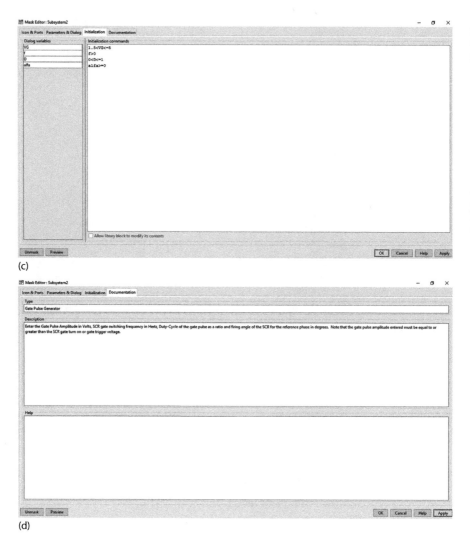

FIGURE 2.7

(c) 'Initialization' menu for gate pulse generator. (d) 'Documentation' menu for gate pulse generator.

entered as 'D*100' and 'Phase delay' is entered as '(alfa*1/(360*f))' s. For Pulse Generator1, connected to the gate of Thyristor1 in Phase Y, the phase delay is entered as '(alfa*1/(360*f)+1/(3*f))', and for Pulse Generator2, connected to Thyristor2 in Phase B, the phase delay is entered as '(alfa*1/(360*f)+2/(3*f))', respectively, with all other entries the same as for the pulse generator connected to the gate of Thyristor in Phase R. The following procedure is used to make the gate pulse generator model interactive.

Fundamentals of Interactive Modelling 21

Step 1: In Figure 2.2, select the 'Pulse Generator', 'Pulse Generator1' and 'Pulse Generator2' modules using the mouse pointer. Then click Diagram → Subsystem and Model Reference → Create Subsystem from the selection to create a subsystem called 'Gate Pulse Generator'. Then, follow Steps 2 to 6 in Section 2.3.2. A view of the 'Parameters and Dialog' menu completed for the gate pulse generator subsystem is shown in Figure 2.7b.

Then, follow Steps 7 to 9 in Section 2.3.2 to open and complete the 'Initialization commands' box. The 'Initialization commands' box completed for the 'Gate Pulse Generator' subsystem is shown in Figure 2.7c.

Finally, follow Steps 10 to 13 given in Section 2.3.2 to complete all the dialogue boxes in the 'Documentation' menu. A view of the 'Documentation' menu completed for the 'Gate Pulse Generator' subsystem is shown in Figure 2.7d.

2.3.5 RLE Load

The RLE load subsystem is shown in Figure 2.4d. This consists of a series-connected resistor, inductor and DC voltage source. The dialogue boxes for the resistor, inductor and DC voltage source are shown in Figure 2.8a–c, respectively. The parameter variable names used for the resistor, inductor and DC voltage source are R1, L1 and E, with units of ohms, henries and volts, respectively.

The following procedure is used to make the RLE load model interactive.

Step 1: In Figure 2.2, select the Resistor, Inductor and DC Voltage source modules using the mouse pointer. Then, click Diagram → Subsystem and Model Reference → Create Subsystem from Selection to create a subsystem called 'RLE Load'. Then, follow Steps 2 to 6 in Section 2.3.2. A view of the 'Parameters and Dialog' menu completed for the RLE load subsystem is shown in Figure 2.8d.

Then, follow Steps 7 to 9 in Section 2.3.2 to open and complete the 'Initialization commands' box. The 'Initialization commands' box completed for the RLE load subsystem is shown in Figure 2.8e.

Finally, follow Steps 10 to 13 given in Section 2.3.2 to complete all the dialogue boxes in the 'Documentation' menu. A view of the 'Documentation' menu completed for the RLE load subsystem is shown in Figure 2.8f.

2.3.6 Output Voltage and Current Measurement

The output voltage and current measurement subsystem is shown in Figure 2.4e. This consists of separate RMS and MEAN(Phasor) blocks to measure the RMS values and average values of the output voltage and load current. The dialogue box for the RMS and MEAN(Phasor) blocks are shown in Figure 2.9a,b respectively. The parameter used is the output frequency 'f' in hertz, which is three times the input AC voltage frequency for both RMS and average value measurement.

22 Power Electronic Converters

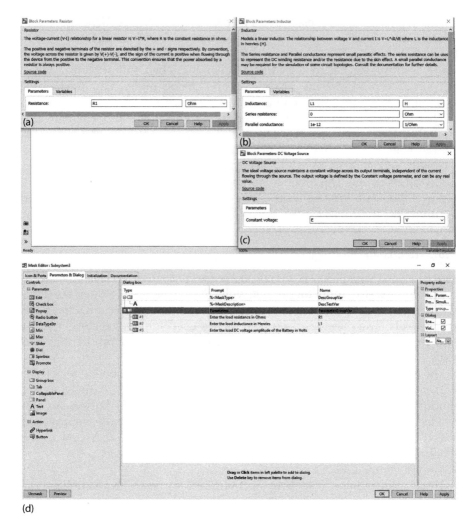

FIGURE 2.8
(a) Parameters for resistor. (b) Parameters for inductor. (c) Parameters for DC voltage source. (d) 'Parameters and Dialog' menu for RLE load.

The following procedure is used to make the output voltage and current measurement model interactive.

Step 1: In Figure 2.2, select the RMS, MEAN(Phasor), RMS1 and MEAN(Phasor)1 modules using the mouse pointer. Then, click Diagram → Subsystem and Model Reference → Create Subsystem from Selection to create a subsystem called 'Output Voltage and Current Measurement'.

Then, follow Steps 2 to 6 in Section 2.3.2. A view of the 'Parameters and Dialog' menu completed for the output voltage and current measurement subsystem is shown in Figure 2.9c.

Fundamentals of Interactive Modelling

(e)

(f)

FIGURE 2.8
(e) 'Intialization' menu for RLE load. (f) 'Documentation' menu for RLE load.

Then, follow Steps 7 to 9 in Section 2.3.2 to open and complete the 'Initialization commands' box. The 'Initialization commands' box completed for the output voltage and current measurement subsystem is shown in Figure 2.9d.

Finally, follow Steps 10 to 13 given in Section 2.3.2 to complete all the dialogue boxes in the 'Documentation' menu. A view of the 'Documentation' menu completed for the output voltage and current measurement subsystem is shown in Figure 2.9e.

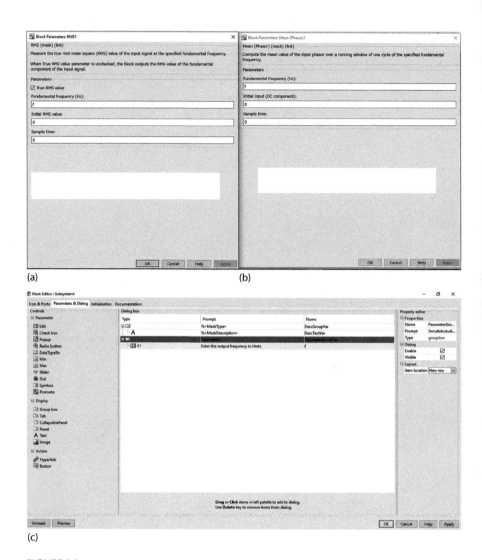

FIGURE 2.9
(a) Parameters for RMS value. (b) Parameters for MEAN(Phasor) value. (c) 'Parameters and Dialog' menu for output voltage and current measurement.

2.3.7 Input Voltage Measurement

The three-phase AC input voltage line-to-neutral RMS value measurement subsystem is shown in Figure 2.4f. This consists of an RMS line-to-neutral input voltage measurement block called RMS2. The parameter used is the input frequency 'f' in hertz. The dialogue box is the same as shown in Figure 2.9a. The following procedure is used to make the input voltage measurement model interactive.

Fundamentals of Interactive Modelling

(d)

(e)

FIGURE 2.9
(d) 'Intialization' menu for output voltage and current measurement. (e) Documentation menu for output voltage and current measurement.

Step 1: In Figure 2.2, select the RMS2 module using the mouse pointer. Then click Diagram → Subsystem and Model Reference → Create Subsystem from Selection to create a subsystem called 'Input Voltage Measurement'. Then, follow Steps 2 to 6 in Section 2.3.2. A view of the 'Parameters and Dialog' menu completed for the input voltage measurement subsystem is shown in Figure 2.10a.

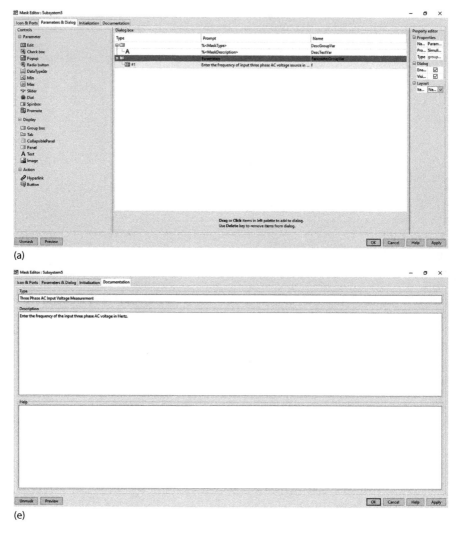

FIGURE 2.10
(a) 'Parameters and Dialog' menu for input voltage measurement. (b) 'Documentation' menu for input voltage measurement.

Then, follow Steps 7 to 9 in Section 2.3.2 to open and complete the 'Initialization commands' box. The 'Initialization commands' box completed for the input voltage measurement subsystem is the same as shown in Figure 2.9d.

Finally, follow Steps 10 to 13 in Section 2.3.2 to complete all the dialogue boxes in the 'Documentation' menu. A view of the 'Documentation' menu completed for the input voltage measurement subsystem is shown in Figure 2.10b.

Fundamentals of Interactive Modelling

2.4 Simulation Results

A simulation of the three-phase half-wave SCR converter model shown in Figure 2.2 was carried out for the parameters shown in Table 2.2. The ode23t(stiff trapezoidal) solver is used [1]. The simulation results for the three-phase AC input voltage are shown in Figure 2.11a–c. The load current and output voltage are shown in Figure 2.12a,b. The simulation results are also tabulated in Table 2.3.

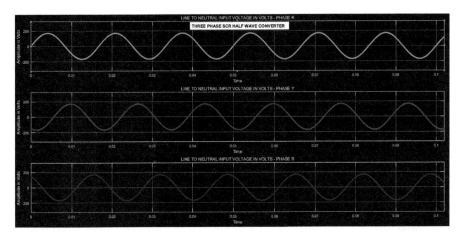

FIGURE 2.11
Three-phase half-wave SCR converter. (a) Line-to-neutral input voltage in volts: Phase R. (b) Line-to-neutral input voltage in volts: Phase Y. (c) Line-to-neutral input voltage in volts: Phase B.

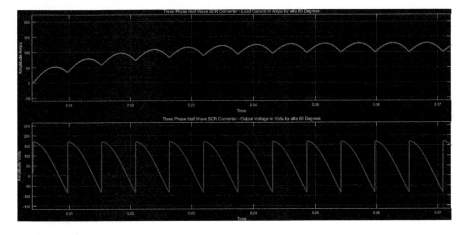

FIGURE 2.12
(a) Three-phase half-wave SCR converter: load current in amps for $\alpha = 90°$. (b) Three-phase half-wave SCR converter: output voltage in volts for $\alpha = 90°$.

TABLE 2.3

Simulation Results for Three-Phase Half-Wave SCR Converter: $\alpha = 90°$

Sl. No.	RMS Line-to-Neutral Input Voltage (V)	Input Frequency (Hz)	RMS Output Voltage (V)	Average Output Voltage (V)	RMS Load Current (A)	Average Load Current (A)
1	120	60	106.7	69.96	120.2	119.9

2.5 Discussion of Results

An interactive model consumes less time compared with a conventional model to simulate and obtain results when there are one or more parameter changes. For example, with a conventional model, if there is a change in the firing angle α, the three pulse generators connected to the gate of the three SCRs have to be opened and the necessary changes in the three dialogue boxes have to be made. This comes to three mouse clicks and three erase and re-enter operations. With an interactive model, it comes to one mouse click and one erase and re-enter operation. Similarly, if there is a change in the input frequency and corresponding thyristor switching frequency, 11 mouse clicks and 11 erase and re-enter operations are required with a conventional model, whereas only four mouse clicks and four erase and re-enter operations are required with an interactive model. For a change in the input line-to-neutral voltage, three mouse clicks and three erase and re-enter operations are required with a conventional model, whereas only one mouse click and one erase and re-enter operation are required with an interactive model. Thus, there is a saving in time, especially when there are a large number of models to be tested for various input parameters. Interactive model development takes more time compared with a conventional model development. But, once an interactive model is developed, the saving in time when testing the model for changes in the input parameters is a definite advantage. If the model topology has one or more capacitors, then variable names starting with C/c, CF/cf, C1/c1, C2/c2 and so on can be given and an interactive model can be developed as for the RLE load subsystem shown in Section 2.3.5.

2.6 Conclusions

Interactive modelling capability is a feature available in the Simulink software. The entire model is developed using the electrical block set in the Simscape foundation library and Semiconductor block set in the power

Fundamentals of Interactive Modelling

systems library. With Simscape and Simulink block set components, S to PS converters and PS to S converters have to be connected. Also, with Simscape components, the Solver Configuration block has to be connected to achieve simulation. The method of developing an interactive model with a three-phase SCR half-wave converter example is presented in this chapter. The method can be easily extended to any power electronic converter topology. With interactive models, the saving in time when testing a power converter for varying parameters is a definite advantage. A virtual power electronics laboratory can be developed with interactive models and, hence, is suitable in an industrial and in a research and development environment.

References

1. Math Works Inc.: "MATLAB/SIMULINK user guide", R2016b, 2016.
2. M.H. Rashid: *SPICE for Power Electronics and Electric Power*, Boca Raton, FL: CRC Press, 2012, pp. 373–377.

3

Interactive Models for AC to DC Converters

3.1 Introduction

In this chapter, the building of interactive models for AC to DC converters, also known as *rectifiers*, is presented. Rectifier circuits can be classified either as half-wave or full-wave bridge rectifiers using semiconductor diodes, or as controlled half-wave or full-wave bridge rectifiers using silicon-controlled rectifiers (SCRs) or thyristors. Modern rectifiers use bipolar junction transistors (BJTs), metal oxide semiconductor field-effect transistors (MOSFETs), insulated gate bipolar transistors (IGBTs) and gate turn-off thyristors (GTOs). Here, interactive system models for single-phase full-wave diode bridge rectifiers (FWDBRs), full-wave controlled bridge rectifiers (FWCBRs) using SCRs and three-phase FWDBRs are developed. The switching function concept is used to develop these models. These models have pop-up menus or dialogue boxes where the required data relating to the AC to DC converter are entered by the user. These system models use Simulink® blocks that solve the governing equations relating to the relevant AC to DC converter. No semiconductor circuit component is used in the model.

3.2 Single-Phase Full-Wave Diode Bridge Rectifier

The single-phase FWDBR circuit schematic is shown in Figure 3.1 [1–4]. The resistive load or RLE load selection using SS1 is shown for convenience only. Let the applied AC voltage V_S be of the form $V_S = V_m * \sin(\omega.t)$. During the positive half-cycle of V_S, diodes D1 and D2 conduct from 0 to π, and during the negative half-cycle, diodes D3 and D4 conduct from π to 2π. The essential derivations of the output voltage are given here for purely resistive load [1].

$$V_{o\_avg} = \frac{1}{\pi} * \int_0^\pi v_m . \sin(\omega.t) . d(\omega.t) = \frac{2.V_m}{\pi} \tag{3.1}$$

31

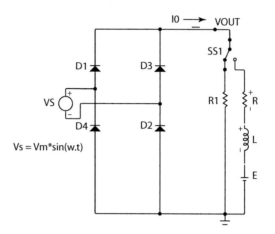

FIGURE 3.1
Single-phase FWDBR.

$$I_{o\_avg} = \frac{V_{o\_avg}}{R} \tag{3.2}$$

$$V_{o\_rms} = \sqrt{\frac{1}{\pi} * \left[\int_0^\pi V_m^2 . \sin^2(\omega t).d(\omega t)\right]} = \frac{V_m}{\sqrt{2}} \tag{3.3}$$

$$I_{o\_rms} = \frac{V_{o\_rms}}{R} \tag{3.4}$$

The output voltage derivations are shown for RLE load [1–4].

$$L.\frac{di_O}{dt} + R.i_O + E = V_m.\sin(\omega t) \tag{3.5}$$

The solution of Equation 3.5 is given here:

$$i_O = \left[\frac{V_m}{(R^2 + \omega^2.L^2)}\right] * \left[R.\sin(\omega t) - \omega L.\cos(\omega t) + \omega L.e^{\frac{-R.t}{L}}\right] - \frac{E.e^{\frac{-R.t}{L}}}{R} + \frac{E}{R} \tag{3.6}$$

3.2.1 Interactive Model for Single-Phase FWDBR with Purely Resistive or with RLE Load

The system model for the single-phase FWDBR is shown in Figure 3.2 [5]. The purely resistive or RLE load selection is combined in one model. The various dialogue boxes are shown in Figure 3.3, where the user can enter

Interactive Models for AC to DC Converters 33

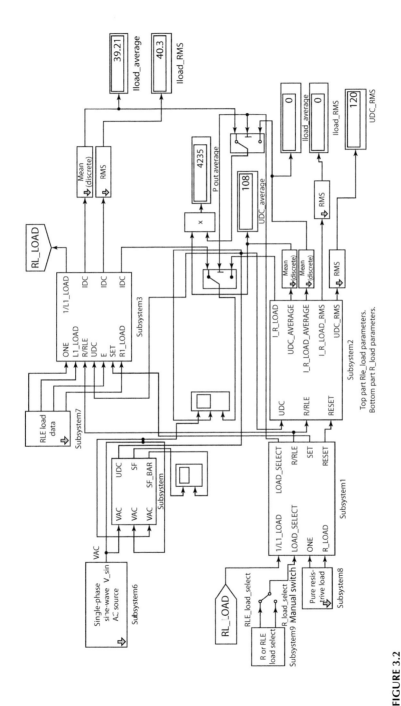

FIGURE 3.2
Single-phase FWDBR.

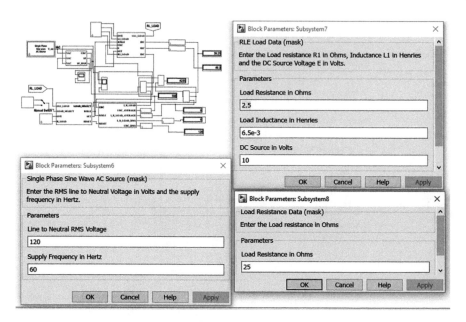

FIGURE 3.3
Single-phase FWDBR with interactive blocks.

the appropriate data required. One model can be used for different system parameters. In Figure 3.2, the selector switch can be double-clicked to select either the purely resistive load or the RLE load option. The value of the line-to-neutral root mean square (RMS) voltage, frequency, the value of RLE loads or that of the purely resistive load used are entered in the appropriate dialogue boxes in Figure 3.3. These values are tabulated in Table 3.1.

The various subsystems of this model are shown in Figure 3.4a–h. These are described as follows.

The single-phase sine-wave AC voltage source in Figure 3.4a is from the 'Sources' block in the 'Simulink' block set. The dialogue box for this is shown in Figure 3.3. The line-to-neutral RMS voltage in volts and the frequency in hertz are entered in this dialogue box. This line-to-neutral RMS voltage and frequency are internally multiplied by 1.414 and 2π, respectively. This

TABLE 3.1

Single-Phase FWDBR Parameters

Sl. No.	Parameter	Value
1	Line-to-neutral RMS voltage	120 V
2	Supply frequency	60 Hz
3	RLE load	2.5 Ω, 6.5e−3 H, 10 V
4	Purely resistive load	25 Ω

Interactive Models for AC to DC Converters

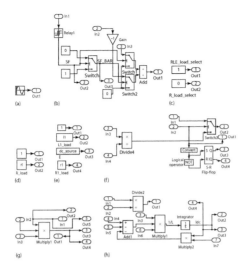

FIGURE 3.4
(a–h) Single-phase FWDBR model subsystems.

sine-wave output is passed to Relay1 in Figure 3.4b. Relay1 is a zero-crossing comparator whose output is the switch function SF, defined in Equation 3.7. The inverse switch function SF_BAR is generated using a switch from the Signal Routing block set, defined in Equation 3.8:

$$SF = 1 \quad \text{for} \quad 0 \leq \omega t \leq \pi$$
$$0 \quad \text{for} \quad \pi \leq \omega t \leq 2\pi \tag{3.7}$$

$$SF\_BAR = 0 \quad \text{for} \quad 0 \leq \omega t \leq \pi$$
$$1 \quad \text{for} \quad 0 \leq \omega t \leq \pi \tag{3.8}$$

The switch generating SF_BAR has 0 and 1 as its first $u(1)$ input and third $u(3)$ input, respectively, while the control or second $u(2)$ input is SF. The switch threshold is 0.5. The switch output is $u(1)$ when input $u(2)$ is greater than or equal to threshold value, else its output is $u(3)$. SF and SF_BAR are passed to the control input $u(2)$ of Switch1 and Switch2, respectively, in Figure 3.4b. The $u(1)$ input to Switch1 is the AC supply voltage V_S, while that for Switch2 is the negative of this supply voltage, $-V_S$. The $u(3)$ inputs to Switch1 and Switch2 are both zero. The threshold values for Switch1 and Switch2 are both 0.5. The Switch1 and Switch2 outputs are $+V_S$ and $-V_S$ when their respective $u(2)$ input is greater than or equal to the threshold value, else their outputs are zero. The output of Switch1 and Switch2 are added using the Add block. The output of this Add block is the rectified DC output voltage U_{dc} (V_{dc}).

Referring to Figure 3.4b, the output voltage of the sum block U_{dc} can be expressed as follows:

$$\left.\begin{aligned} U_{dc} &= SF.\left(V_S\right) + SF\_BAR.\left(-V_S\right) \\ &= V_S \quad \text{for} \quad 0 \le \omega t \le \pi \\ &= -V_S \quad \text{for} \quad \pi \le \omega t \le 2\pi \end{aligned}\right] \tag{3.9}$$

In Equation 3.9, V_S is the sine-wave input voltage.

The type of load select module shown in Figure 3.4c has two constants, 1 and 0, as input to the single pole double throw (SPDT) switch. Double-clicking the SPDT switch in Figure 3.2 selects either the purely resistive load or the RLE load.

Figure 3.4d, e corresponds to the dialogue box for load resistance data and RLE load data, respectively, shown in Figure 3.3.

If the SPDT switch is thrown to 0 in Figure 3.2, then the purely resistive load R_LOAD is selected. Referring to Figure 3.4f, it is seen that the Divide4 block calculates 1/(R_LOAD), which is passed to the $u(3)$ input of Switch3. The 1/(L1_LOAD) input from Figure 3.4h is passed to the $u(1)$ input of Switch3. The control input $u(2)$ for Switch3 is from the SPDT switch connected to the output of Figure 3.4c. The threshold for Switch3 is 0.5, and output is either 1/(L1_LOAD) or 1/(R_LOAD), depending on whether the $u(2)$ input is 1 or 0, respectively. The $u(2)$ input of Switch3 is also connected to the S input and through a NOT gate to the R input of the S–R flip-flop, as shown in Figure 3.4f. If the SPDT switch is thrown to 0 or 1, the S–R flip-flop enables the bottom display or top display, respectively, in Figure 3.2.

With SPDT in the 0 or bottom position, the value 1/(R_LOAD) is selected. In Figure 3.4g, the DC output voltage U_{dc} is multiplied by 1/(R_LOAD) and by RESET, which is 1, using the Multiplier1 block to obtain the DC load current I_{dc}.

If the SPDT switch is thrown to 1, then the RLE load in Figure 3.4e is selected. Referring to Figure 3.4h, it is seen that the Divide2 block calculates 1/(L1_LOAD), which is applied to the $u(1)$ input of Switch3 in Figure 3.4f. Referring to Figure 3.4f, the $u(2)$ input of this Switch3 is 1, which is greater than its threshold of 0.5 and selects output 1/(L1_LOAD). This $u(2)$ input also sets the S–R flip-flop and Q is HIGH and enables the top display for the RLE load.

In Figure 3.4h, the Add1, Multiply1, Multiply2 and Integrator blocks solve Equation 3.5 and the resulting output of the Integrator is the DC load current I_{dc} with RLE load. The S–R latch enables the top display in Figure 3.2. The RMS and average value of load voltage U_{dc} and the mean and RMS value of the load current and the average power output are displayed.

3.2.2 Simulation Results

A simulation of the model for single-phase FWDBR using the ode23t(mod. stiff/trapezoidal) solver for data given in Table 3.1 for the purely resistive load and the RLE load was conducted. The simulation results for purely

Interactive Models for AC to DC Converters 37

resistive load are shown in Figure 3.5a–c and those for RLE load are shown in Figure 3.6a–c. The meter readings observed are tabulated in Table 3.2 for both cases.

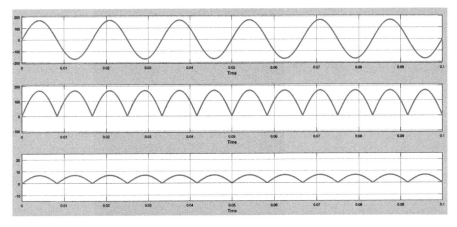

FIGURE 3.5
(a) Single-phase FWDBR with purely resistive load-source voltage volts. (b) Output voltage volts. (c) Load current in amps.

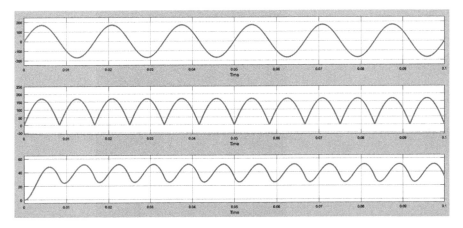

FIGURE 3.6
(a) Single-phase FWDBR with RLE load-source voltage volts. (b) Output voltage volts. (c) Load current in amps.

TABLE 3.2

Simulation Results for Single-Phase FWDBR

Sl. No.	Type of Load	Udc(RMS) (V)	Udc(AVG) (V)	I_load(RMS) (A)	I_load(AVG) (A)	P_out (W)
1	R only	120	108	4.799	4.321	466.8
2	RLE	120	108	40.3	39.21	4235

3.3 Single-Phase Full-Wave SCR Bridge Rectifier

The single-phase full-wave SCR bridge rectifier is also known as the *single-phase full-wave controlled bridge rectifier* (FWCBR) [1–4]. This is shown in Figure 3.7. Assuming pure resistive load, during the positive half-cycle of V_S, thyristors T1 and T2 are fired at an angle α and conduct from α to π. During the negative half-cycle of V_S, thyristors T3 and T4 are fired at angle $(\pi+\alpha)$ and conduct from $(\pi+\alpha)$ to 2π. Thus, by varying the firing angle α, the average voltage across the load can be varied. The system model for this FWCBR is developed combining the purely resistive load and RLE load in one model. For the RLE load, continuous load current conduction is assumed. In this case, with the RLE load, thyristors T1 and T2 conduct from α to $(\pi+\alpha)$ when fired at an angle α during the positive half-cycle of V_S and thyristors T3 and T4 conduct from $(\pi+\alpha)$ to $(2\pi+\alpha)$, when fired at an angle $(\pi+\alpha)$ during the negative half-cycle of V_S. This model can be used for continuous conduction mode only with RLE load.

The essential derivation for output voltage and load current is shown here for purely resistive load.

$$V_{dc} = \frac{1}{\pi} * \left[\int_{\alpha}^{\pi} V_m \cdot \sin(\omega t) \cdot d(\omega t) \right] = \frac{V_m}{\pi} * \left[1 + \cos(\alpha) \right] \quad (3.10)$$

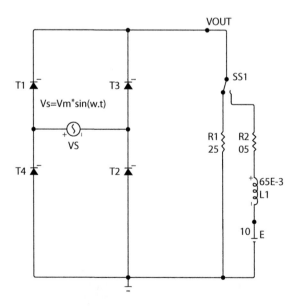

FIGURE 3.7
Single-phase full-wave SCR bridge rectifier.

Interactive Models for AC to DC Converters

$$V_{rms} = \sqrt{\frac{1}{\pi}*\left[\int_{\alpha}^{\pi}V_m^2.\sin^2\left(\omega t\right).d\left(\omega t\right)\right]} = \frac{V_m}{\sqrt{2}}*\left[\sqrt{1-\frac{\alpha}{\pi}+\frac{\sin\left(2\alpha\right)}{2\pi}}\right] \quad (3.11)$$

$$\text{Average input power } P_{in} = \frac{1}{\pi.R}\left[\int_{\alpha}^{\pi}\left(V_m.\sin\left(\omega t\right)\right)^2.d\left(\omega t\right)\right] \quad (3.12)$$

$$= \frac{V_m^2}{2R}*\left[1-\frac{\alpha}{\pi}+\frac{\sin\left(2\alpha\right)}{2\pi}\right] \quad (3.13)$$

The essential derivation for output voltage and load current is shown here for RLE load [1].

$$V_{dc} = \frac{1}{\pi}*\left[\int_{\alpha}^{\pi+\alpha}V_m.\sin(\omega t).d(\omega t)\right] = \frac{2V_m}{\pi}.\cos(\alpha) \quad (3.14)$$

$$V_{rms} = \sqrt{\frac{1}{\pi}*\left[\int_{\alpha}^{\pi+\alpha}V_m^2.\sin^2(\omega t).d(\omega t)\right]} = \frac{V_m}{\sqrt{2}} \quad (3.15)$$

$$L.\frac{di_L}{dt}+R.i_L+E = V_m.\sin(\omega t) \quad for \quad i_L \succ 0 \quad (3.16)$$

The solution gives

$$i_L = \frac{V_m}{Z}*\sin(\omega t-\theta)-\frac{E}{R}+\left[I_{LO}+\frac{E}{R}-\frac{V_m}{Z}*\sin(\alpha-\theta)\right]*e^{(R/L)*(\alpha/\omega-t)} \quad (3.17)$$

where $I_{LO}=I_{LI}$ and θ are given by

$$I_{LO} = I_{L1} = \frac{V_m-\sin(\alpha-\theta)-\sin(\alpha-\theta)*e^{-(R/L)*(\pi)/\omega}}{Z*\left(1-e^{-(R/L)*(\pi/\omega)}\right)}-\frac{E}{R} \quad for \quad I_{LO} \succ 0$$

$$\theta = \tan^{-1}\left(\frac{\omega*L}{R}\right)$$

$$(3.18)$$

3.3.1 Model for Single-Phase FWCBR with Purely Resistive or with RLE Load

The system model for the single-phase FWCBR is shown in Figure 3.8 [5]. The purely resistive or RLE load selection is combined in one model. The various dialogue boxes are shown in Figure 3.9, where the user can enter

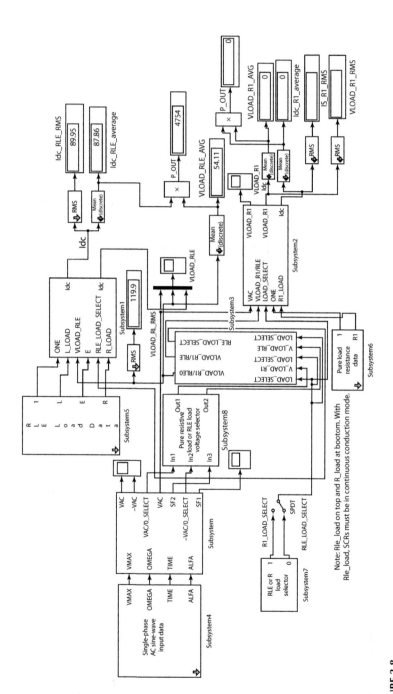

FIGURE 3.8
Single-phase full-wave SCR bridge rectifier.

Interactive Models for AC to DC Converters 41

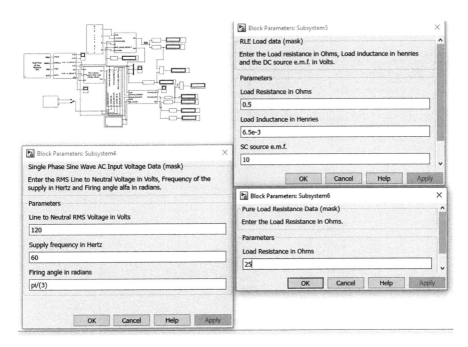

FIGURE 3.9
Single-phase full-wave SCR bridge rectifier with interactive blocks.

the appropriate data required. One model can be used for different system parameters. In Figure 3.8, the selector switch can be double-clicked to select either the purely resistive load or the RLE load option. The value of the line-to-neutral RMS voltage, the frequency, the value of RLE loads used and/or that of the purely resistive load used are entered in the appropriate dialogue boxes in Figure 3.9. These values are tabulated in Table 3.3. The various subsystems of this model are shown in Figure 3.10a–i. These are described next.

The blocks in Figure 3.10a, along with the mux and Fcn block in Figure 3.10b, generate the single-phase sine-wave AC input voltage V_S. The dialogue box for this is shown in Figure 3.9. The line-to-neutral RMS voltage (in volts), the frequency (in hertz) and the firing angle α are entered in this dialogue box. The line-to-neutral RMS voltage and frequency are internally multiplied by 1.414 and 2π respectively. Another Fcn1 block in Figure 3.10b generates the delayed sine-wave lagging by firing angle α with respect to the input sine wave. This delayed or lagging sine wave is passed to relay, which generates switch function SF1. The sine-wave input voltage V_S from the Fcn block is passed to Relay1, which generates another switch function SF2. The switch functions SF1 and SF2 are defined in Equations 3.19 and 3.20.

$$SF1 = \begin{cases} 1 & \text{for} \quad (\alpha) \leq \omega.t \leq (\pi + \alpha) \\ 0 & \text{for} \quad (\pi + \alpha) \leq \omega.t \leq (2\pi + \alpha) \end{cases} \quad (3.19)$$

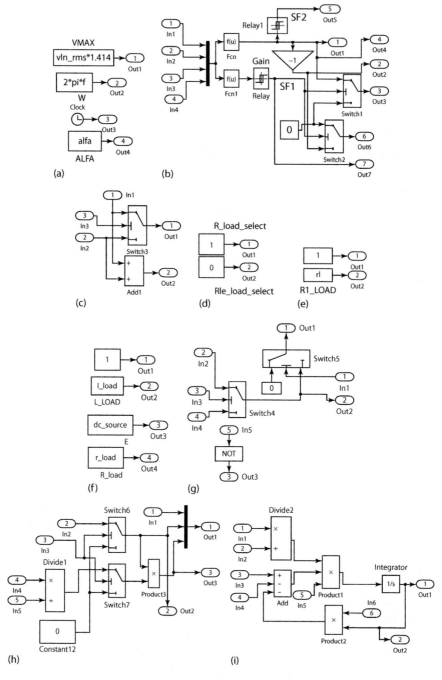

FIGURE 3.10
(a–i) Single-phase FWCBR model subsystems.

Interactive Models for AC to DC Converters 43

$$SF2 = \begin{cases} 1 & \text{for} \quad (0) \leq \omega.t \leq (\pi) \\ 0 & \text{for} \quad (\pi) \leq \omega.t \leq (2\pi) \end{cases} \tag{3.20}$$

Both Relay and Relay1 are zero-crossing comparators whose output is HIGH or logic 1 when the respective sine-wave input crosses zero and becomes positive, and their output is LOW or logic 0 when the respective sine-wave input crosses zero and goes negative. Switch1 and Switch2 have $u(1)$ input $+V_S$ and 0, $u(3)$ input 0 and $-V_S$, respectively, and the $u(2)$ input for both is SF1. Both switches have a threshold value of 0.5. Switches 1 and 2 output $+V_S$ and 0 when their respective $u(2)$ input is greater than or equal to the threshold value, else their outputs are 0 and $-V_S$, respectively. Referring to the mode of connection of Switch1 and Switch2 in Figure 3.10b, the respective outputs of Switch1 and 2 can be mathematically expressed as follows:

$$\text{Output of Switch1 Out3} = SF1.V_S \tag{3.21}$$

$$\text{Output of Switch2 Out6} = (1 - SF1).(-V_S) \tag{3.22}$$

Switch1 and 2 output and SF2 are passed to the $u(1)$, $u(3)$ and $u(2)$ input of Switch3, respectively, in Figure 3.10c. The threshold for Switch3 is 0.5 and output Switch1 value if $u(2)$ input SF2 is greater than or equal to this threshold value, else output Switch2 value. The output of Switch3 in Figure 3.10c can be expressed mathematically as follows:

$$\text{Output of Switch3 Out1} = SF2.(SF1.V_S) + (1 - SF2).(1 - SF1).(-V_S) \tag{3.23}$$

The output of Switch3 can be expressed as follows from Equations 3.19, 3.20 and 3.23:

$$\begin{aligned} \text{Output of Switch3 Out1} &= +V_S \quad \text{for} \quad \alpha \leq \omega.t \leq \pi \\ &= -V_S \quad \text{for} \quad (\pi + \alpha) \leq \omega.t \leq 2\pi \end{aligned} \tag{3.24}$$

Equation 3.24 represents the output voltage of single-phase FWCBR with purely resistive load. In Figure 3.10c, the output of the Add1 block can be expressed as follows:

$$\begin{aligned} \text{Output of Add1 block Out2} &= +V_S \quad \text{for} \quad \alpha \leq \omega.t \prec (\pi + \alpha) \\ &= -V_S \quad \text{for} \quad (\pi + \alpha) \leq \omega.t \leq (2\pi + \alpha) \end{aligned} \tag{3.25}$$

Equation 3.25 represents the output voltage of the single-phase FWCBR with RLE load in the *continuous conduction mode*.

The type of load select module shown in Figure 3.10d has two constants, 1 and 0, as input to the SPDT switch. Double-clicking the SPDT switch in Figure 3.8 selects either the purely resistive load or the RLE load.

Figure 3.10e, f corresponds to the dialogue box for pure load resistance data and RLE load data, respectively, shown in Figure 3.9.

If the SPDT switch is thrown to 1, then purely resistive load is selected. The Switch3 and Add1 block output of Figure 3.10c and the SPDT output of Figure 3.10d are applied to the $u(1)$, $u(3)$ and $u(2)$ input of Switch4 in Figure 3.10g. Switch4 has a threshold value of 0.5, and selects the purely resistive load voltage or RLE load voltage as output depending on its $u(2)$ input corresponding to whether the SPDT switch is 1 or 0. This Switch4 output and SPDT output are passed as $u(3)$ and $u(2)$ input to Switch5, whose $u(1)$ input is 0. Switch5 has a threshold value of 0.5 and its output is 0 if its $u(2)$ input corresponding to the SPDT switch is 1, and outputs the Switch4 value when its $u(2)$ input is 0. Thus, the Switch5 output is always the load voltage corresponding to the RLE load. The Switch4 output can be either the purely resistive or the RLE load voltage, depending on the position of the SPDT switch. The SPDT switch output is inverted using a NOT gate to enable selection of the RLE load.

Referring to Figure 3.10h, the Divide1 block calculates 1/(R1_LOAD), which is passed to the $u(1)$ input of Switch7, whose $u(3)$ input is 0, and the $u(2)$ input is from the SPDT switch output. The threshold is 0.5, and Switch7 output is either 1/(R1_LOAD) or 0, depending on whether the $u(2)$ input corresponding to the SPDT switch is 1 or 0. The Switch4 output from Figure 3.10g is passed to the $u(1)$ input of Switch6, whose $u(3)$ input is 0, and the $u(2)$ input is the output of the SPDT switch. The threshold for Switch6 is 0.5, and the output is the purely resistive load voltage or 0, depending on whether the $u(2)$ input corresponding to the SPDT switch output is 1 or 0. The Product3 block multiplies the Switch6 and Switch7 outputs, giving a DC load current with a purely resistive load.

RLE load is selected by throwing SPDT to 0. Referring to Figure 3.10f, g and i, it is seen that Switch4 and Switch5 output the load voltage across the RLE load. The Divide2 block calculates 1/(L_LOAD). The Add block, along with the Divide2, Product1, Product2 and Integrator blocks, solves Equation 3.16 to find the load current through the RLE load. Using the NOT gate in Figure 3.10g connected to the SPDT switch, the upper indicating meters are enabled for the RLE load and the lower indicating meters are enabled for the purely resistive load.

3.3.2 Simulation Results

The simulation of the single-phase FWCBR model was developed using the ode23t(mod. stiff/trapezoidal) solver for the data given in Table 3.3, for purely resistive and RLE loads. The simulation results for the purely resistive load are shown in Figure 3.11a–c and those for the RLE load are shown in Figure 3.12a–c. The meter readings observed are tabulated in Table 3.4 for both cases.

Interactive Models for AC to DC Converters   45

TABLE 3.3

Single-Phase FWCBR Simulation Parameters

Sl. No.	Parameter	Value
1	Line-to-neutral RMS voltage	120 V
2	Supply frequency	60 Hz
3	RLE load	0.5 Ω, 6.5e−3 H, 10 V
4	Purely resistive load	25 Ω
5	Firing angle α	π/3 rad

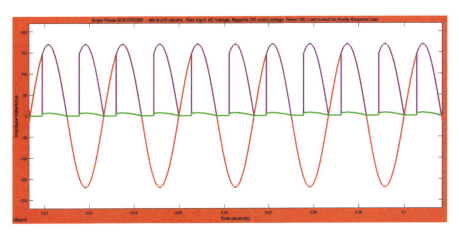

FIGURE 3.11
Single-phase FWCBR. Alfa is pi/3 radians. Red, input AC voltage; magenta, DC output voltage; green, DC load current for purely resistive load.

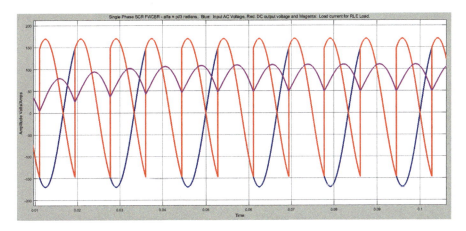

FIGURE 3.12
Single-phase FWCBR. Alfa is pi/3 radians. Blue, input AC voltage; red, DC output voltage; magenta, load current for RLE load.

TABLE 3.4

Single-Phase FWCBR: Simulation Results for $\alpha = \pi/3$ rad

Sl. No.	Type of Load	Vdc(RMS) (V)	Vdc(AVG) (V)	I_load(RMS) (A)	I_load(AVG) (A)	P_out (W)
1	R only	107.5	81.07	4.301	3.243	262.9
2	RLE	120	54.11	90.06	88.01	4762

3.4 Three-Phase Full-Wave Diode Bridge Rectifier

The three-phase FWDBR schematic is shown in Figure 3.13. It is used for high-power applications. The diodes are numbered in conduction sequences and each diode conducts 120°. The conduction sequences for the diodes are: D1–D2, D3–D2, D3–D4, D5–D4, D5–D6, D1–D6, D1–D2 and so on, giving six pulses of ripples at the output voltage. The pair of diodes that are connected between that pair of lines having the highest amount of instantaneous line-to-line voltage will conduct [1–4].

The essential derivation for the purely resistive load is shown here [1–4]:

$$\left. \begin{aligned} V_{an} &= V_m \cdot \sin(\omega t) \\ V_{bn} &= V_m \cdot \sin(\omega t - 120) \\ V_{cn} &= V_m \cdot \sin(\omega t - 240) \end{aligned} \right| \tag{3.26}$$

$$\left. \begin{aligned} V_{ab} &= \sqrt{3} \cdot V_m \cdot \sin(\omega t + 30) \\ V_{bc} &= \sqrt{3} \cdot V_m \cdot \sin(\omega t - 90) \\ V_{ca} &= \sqrt{3} \cdot V_m \cdot \sin(\omega t - 210) \end{aligned} \right| \tag{3.27}$$

The average output voltage V_{dc} is derived as follows [1–4]:

$$
\begin{aligned}
V_{dc} &= \frac{6}{\pi} * \int_{\pi/6}^{\pi/2} V_m \cdot \sin(\omega t) \cdot d(\omega t) \\
&= \frac{6}{\pi} * \int_{0}^{\pi/6} \sqrt{3} * V_m \cdot \cos(\omega t) \cdot d(\omega t) \\
&= \frac{3\sqrt{3} \cdot V_m}{\pi} = 1.654 * V_m
\end{aligned}
\tag{3.28}
$$

Interactive Models for AC to DC Converters

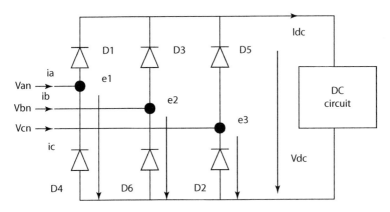

FIGURE 3.13
Three-phase FWDBR.

The RMS output voltage V_{rms} is derived as follows [1–4]:

$$V_{rms} = \frac{6}{\pi} * \int_{\pi/6}^{\pi/2} V_m^2 * \sin^2(\omega t).d(\omega t)$$

$$= \frac{6}{\pi} * \int_{0}^{\pi/6} 3V_m^2 * \cos^2(\omega t).d(\omega t) \quad (3.29)$$

$$= V_m * \sqrt{\left[\frac{3}{2} + \frac{9\sqrt{3}}{4\pi}\right]} = 1.6554 * V_m$$

If the load resistance is purely resistive, the peak or maximum current through the diode is $I_m = \sqrt{3}.V_m/R$, where R is the load resistance. The RMS value of the diode current is derived as follows:

$$I_r = \sqrt{\left[\frac{4}{2\pi} * \int_{0}^{\pi/6} I_m^2 * \cos^2(\omega t).d(\omega t)\right]} \quad (3.30)$$

$$= 0.5518 * I_m$$

The RMS value of the supply current or line current is derived as follows:

$$I_S = \sqrt{\left[\frac{8}{2\pi} * \int_{0}^{\pi/6} I_m^2 * \cos^2(\omega t).d(\omega t)\right]} \quad (3.31)$$

$$= 0.7804 * I_m$$

where I_m is the peak value of the three-phase line current (in Equations 3.30 and 3.31).

If L_s is the source inductance in each of the three phases and f is the supply frequency, then the average reduction in output voltage due to source or commutating inductances is given as follows [1–4]:

$$V_X = 6.f.L_s.I_{dc} \tag{3.32}$$

where I_{dc} is the load current.

3.4.1 Model for Three-Phase FWDBR with Purely Resistive Load

Before describing the subsystems in detail, the development of the system model for the three-phase FWDBR is explained as follows:

The system model for the three-phase FWDBR is developed based on reference [6]. The Heaviside function gk for $k = 1, 2, 3, \ldots$ is defined in Figure 3.14. This Heaviside function determines whether the diode is conducting or in the blocking state [6]. In Figure 3.13, $e1$, $e2$ and $e3$ are the arm voltages and $u1$, $u2$ and $u3$ are the phase voltages. The relations connecting them are given here:

$$e1 = g1 * V_{dc} \tag{3.33}$$

$$e2 = g2 * V_{dc} \tag{3.34}$$

$$e3 = g3 * V_{dc} \tag{3.35}$$

$$u1 = f1 * V_{dc} \tag{3.36}$$

$$u2 = f2 * V_{dc} \tag{3.37}$$

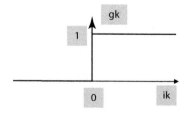

FIGURE 3.14
Definition of the gk function ($k = 1,2,3,\ldots$).

Interactive Models for AC to DC Converters 49

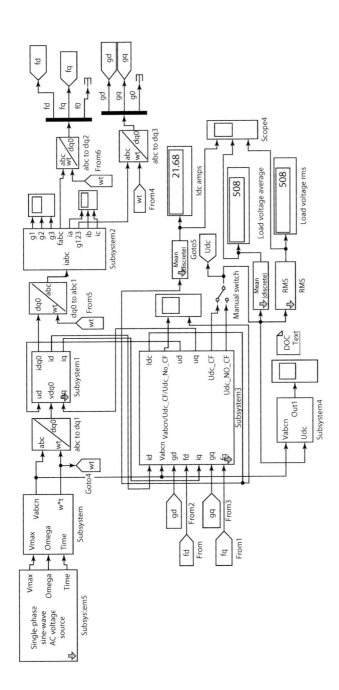

FIGURE 3.15
Three-phase FWDBR.

$$u3 = f3 * V_{dc} \tag{3.38}$$

where functions $f1$, $f2$ and $f3$ are defined as follows:

$$f1 = \frac{2g1 - g2 - g3}{3} \tag{3.39}$$

$$f2 = \frac{2g2 - g3 - g1}{3} \tag{3.40}$$

$$f3 = \frac{2g3 - g1 - g1}{1} \tag{3.41}$$

The DC current is given here:

$$I_{dc} = g1.i1 + g2.i2 + g3.i3$$

If R_f and L_f are the resistance and inductance of each phase of the AC mains, then $e1$, $e2$ and $e3$ are related as follows:

$$e1 = R_f.i_a + L_f.\frac{di_a}{dt} + u1 \tag{3.43}$$

$$e2 = R_f.i_b + L_f.\frac{di_b}{dt} + u2 \tag{3.44}$$

$$e3 = R_f.i_c + L_f.\frac{di_c}{dt} + u3 \tag{3.45}$$

To simplify modelling and procedure, the abc to dq and dq to abc transformation is defined as given here:

$$\begin{bmatrix} x_1 \\ x_2 \\ x_3 \end{bmatrix} = \begin{bmatrix} \cos(\theta) & \sin(\theta) \\ \cos(\theta - 2\pi/3) & \sin(\theta - 2\pi/3) \\ \cos(\theta + 2\pi/3) & \sin(\theta + 2\pi/3) \end{bmatrix} * \begin{bmatrix} x_d \\ x_q \end{bmatrix} \tag{3.46}$$

$$\begin{bmatrix} x_d \\ x_q \end{bmatrix} = \begin{bmatrix} \cos(\theta) & \cos(\theta - 2\pi/3) & \cos(\theta + 2\pi/3) \\ \sin(\theta) & \sin(\theta - 2\pi/3) & \sin(\theta + 2\pi/3) \end{bmatrix} * \begin{bmatrix} x_1 \\ x_2 \\ x_3 \end{bmatrix} \tag{3.47}$$

In Equations 3.46 and 3.47, x may be a voltage, current or Heaviside function. The dq axis voltages and load current I_{dc} are defined as follows:

Interactive Models for AC to DC Converters 51

$$I_{dc} = gd.id + gq.iq \tag{3.48}$$

$$v_d = R_f.id + L_f.\frac{did}{dt} + ud \tag{3.49}$$

$$v_q = R_f.iq + L_f.\frac{diq}{dt} + uq \tag{3.50}$$

Equations 3.33 to 3.50 are used to develop the system model for the three-phase FWDBR.

The system model for the three-phase FWDBR is shown in Figure 3.15 [5, 6]. The various user dialogue boxes are shown in Figure 3.16. The user can enter the required data in the appropriate dialogue box. This model is suitable only for the purely resistive load. The various subsystems are shown in Figure 3.17a–f and are described as follows.

Figure 3.17a corresponds to the dialogue box in Figure 3.15, where the line-to-neutral RMS voltage in volts and frequency in hertz are entered. Figure 3.17b generates three-phase AC using the three-input mux and the three Fcn blocks. The *abc* to *dq0* and *dq0* to *abc* blocks are from the Power Systems Transformations block set, which is used for the axis transformation

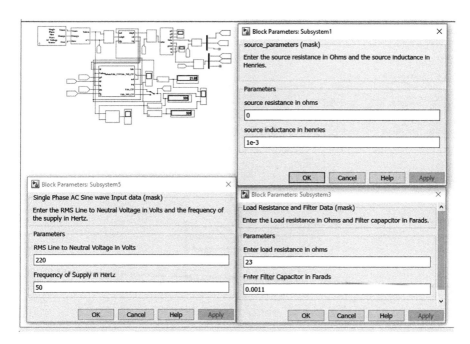

FIGURE 3.16
Three-phase FWDBR with interactive blocks.

of voltage, current and Heaviside functions *gk* and *fk*. Figure 3.17c solves the equation for the *dq* axis line currents given by Equations 3.49 and 3.50. In Figure 3.17c, *rs* and *ls* represent the AC line resistance and inductance in each phase, corresponding to R_f and L_f in Equations 3.49 and 3.50. Figure 3.17c opens up the source parameters dialogue box shown in Figure 3.16. The *dq* axis line current is transformed to the *abc* axis line currents using the *dq*0 to *abc* block. Figure 3.17d has the *abc* axis line currents as inputs and generates the Heaviside functions *g*1, *g*2 and *g*3 for the three phases as defined in

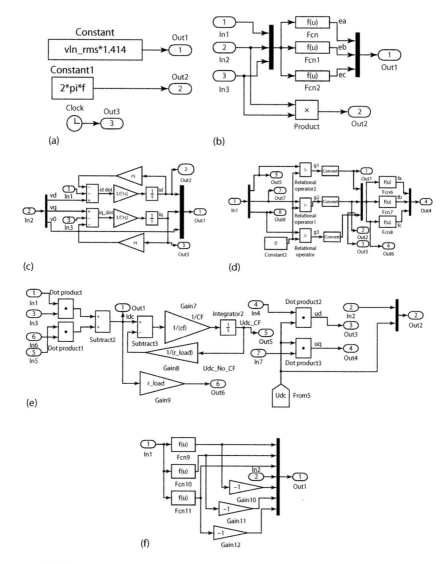

FIGURE 3.17
Three-phase FWDBR model subsystems.

Interactive Models for AC to DC Converters

Figure 3.14, and then using g1, g2 and g3, the functions fa, fb and fc are generated as defined by Equations 3.39 to 3.41, where f1, f2 and f3 correspond to fa, fb and fc. Heaviside functions g1, g2, g3 and f1, f2, f3 are transformed to their corresponding dq axis equivalent using abc to dq0 blocks as shown in Figure 3.15. Figure 3.17e opens the 'Load Resistance and Filter Data' dialogue box, which is shown in Figure 3.16. In Figure 3.17e, load current I_{dc} is calculated using Equation 3.48 using inputs id, gd, iq, gq, two dot product multipliers and an Add block.

This value of I_{dc} is used to calculate $I_{dc}*R$ using the Gain9 block to find the load voltage V_{dc}(Udc_NO_CF) without filter, where R is the load resistance. With the capacitor filter C_F in parallel to the load resistance, V_{dc}(Udc_CF) is calculated by calculating the capacitor filter current I_C as follows:

$$I_C = I_{dc} - (V_{dc}/R)$$
$$V_{dc} = \frac{1}{C_F}\int I_C.dt \qquad (3.51)$$

Equation 3.51 is solved using the Subtract3, Gain7, Gain8 and Integrator2 modules in Figure 3.17e. Further, Figure 3.17e is used to find the dq0 axis phase voltages ud and uq using V_{dc}(Udc) and fd, fq as inputs and two dot product multiplier blocks. Figure 3.17f is used to find the three line-to-line voltages and their inverted values using three Fcn blocks and three gain blocks.

In Figure 3.15, the DC output voltage U_{dc} can be selected with or without the capacitor filter by using the SPDT marked Manual Switch.

The values of the RMS line-to-neutral voltage in volts, frequency in hertz, source resistance in ohms, source inductance in henries, load resistance in ohms and filter capacitor in farads are entered in the appropriate dialogue box in Figure 3.16. These values are shown in Table 3.5 [6].

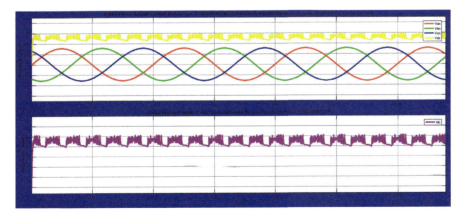

FIGURE 3.18
Three-phase FWDBR. (a–d) Three-phase line-to-neutral input voltages and DC output voltage with no filter capacitor. (e) Average load current Idc amps with no filter capacitor.

TABLE 3.5
Three-Phase FWDBR: Parameters

Sl. No	Parameters	Value
1	Line-to-neutral RMS voltage	220 V
2	Supply frequency	50 Hz
3	Source resistance	0 Ω
4	Source inductance	1e–3 H
5	Load resistance	23 Ω
6	Filter capacitance	0.0011 F

3.4.2 Simulation Results

The simulation of the three-phase FWDBR was carried out using the ode1(Euler) fixed-step solver for the purely resistive load, first by excluding the filter capacitor from the circuit and then by including the filter capacitor, for the data shown in Table 3.5. The simulation results of input voltage, output voltage and average load current excluding the filter capacitor from the circuit are shown in Figure 3.18a–e and Figure 3.19a–g. The simulation results for these including the filter capacitor are shown in Figure 3.20a–e and Figure 3.21a–g. The values of the RMS, average load voltage and average load current observed for both cases by simulation are recorded in Table 3.6. The average and RMS value of load voltage by calculation using Equations 3.28 and 3.29 yield 514.5 and 515 V respectively. The reduction in the output voltage due to source inductance of 1 mH was calculated using Equation 3.32 and the net value of the output voltage was found to be 507.9 V. The average load current calculated using the formula comes to 22.37 A. The respective

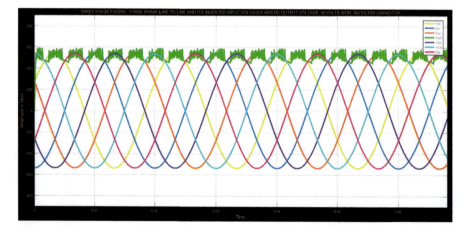

FIGURE 3.19
Three-phase FWDBR. Three-phase line to line and its inverted input voltages and DC output voltage in volts with no filter capacitor.

Interactive Models for AC to DC Converters 55

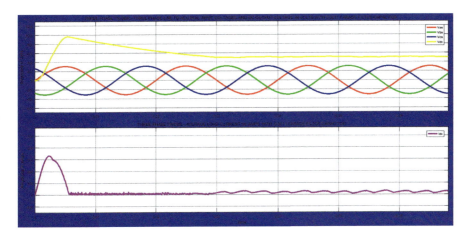

FIGURE 3.20
Three-phase FWDBR. (a–d) Three-phase line-to-neutral input voltages and DC output voltage in volts with 0.0011 farads filter capacitor. (e) Average load current Idc amps with 0.0011 farads capacitor.

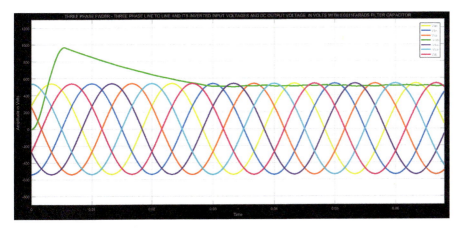

FIGURE 3.21
Three-phase FWDBR. Three-phase line to line and its inverted input voltages and DC output voltage in volts with 0.0011 farads filter capacitor.

TABLE 3.6
Simulation Results for Three-Phase FWDBR

Sl. No.	RMS Load Voltage (V)	Average Load Voltage (V)	Average Load Current (A)	Remarks
1	511	509.6	22.15	No filter capacitor
2	508	508	21.68	With filter capacitor of 0.0011 farads

values recorded in Table 3.6 by simulation closely agree with the theoretically calculated values [6].

3.5 Conclusions

The interactive model for the single-phase FWDBR based on the equations describing the system is presented. One model is suitable for either a purely resistive load or an RLE load. The type of load is easily selected using the selector switch. The simulation results in Figures 3.5 and 3.6 and the values in Table 3.2 agree well with reference [1]. The interactive model for the single-phase FWCBR based on the equations describing the system is presented. One model is suitable for either a purely resistive load or an RLE load. The type of load is easily selected using the selector switch. The simulation results in Figures 3.11 and 3.12 and the calculated values in Table 3.4 agree well with reference [1]. A novel method of system modelling the three-phase FWDBR using the Heaviside function is presented. The method is very similar to the one discussed in reference [6], except that here abc to dq and dq to abc transformations are used instead of abc to $\alpha\beta$ and $\alpha\beta$ to abc transformations. The recorded values of the average and RMS load voltage and load current in Table 3.6 by simulation agree closely with the corresponding values calculated using the formula.

References

1. M.H. Rashid: *Power Electronics Circuits, Devices and Applications*, Upper Saddle River, NJ: Pearson Education, Pearson Prentice Hall, 2004.
2. I. Batarseh: *Power Electronic Circuits*, Hoboken, NJ: Wiley, 2004.
3. D.W. Hart: *Introduction to Power Electronics*, Upper Saddle River, NJ: Prentice Hall, 1997.
4. N. Mohan, T.M. Undeland, and W.P. Robbins: *Power Electronics: Converters, Applications and Design*, Hoboken, NJ: Wiley, 1995.
5. N.P.R. Iyer: "Matlab/Simulink modules for modelling and simulation of power electronic converters and electric drives", M.E. by research thesis, University of Technology Sydney, NSW, 2006, Chapter 3, pp. 22–58.
6. G.D. Marques: "A simple and accurate system simulation of three-phase diode rectifiers"; *IEEE-IECON*; Aachen, Germany, 1998; pp. 416–421.

4

Interactive Models for DC to AC Converters

4.1 Introduction

In this chapter, the building of interactive system models for DC to AC converters, also known as *inverters*, is presented. Inverter circuits use silicon-controlled rectifiers (SCRs), BJTs, MOSFETs, IGBTs and GTOs. Here, initially, circuit models for three-phase 180° mode and 120° mode inverters using semiconductor switches such as BJTs are presented. Then, the switching function concept is used to develop system models for the inverter for these two modes of gate drive. These interactive models have pop-up menus or dialogue boxes where the required data relating to the DC to AC converter are entered by the user. These system models use modules from the Simulink® block set, which solve the governing equations relating to the relevant DC to AC converter. Here, the governing equations are the gate drive pattern for the two modes of inverter operation. No semiconductor circuit component is used in the system model.

4.2 Three-Phase 180° Mode Inverter

The three-phase 180° mode inverter topology is shown in Figure 4.1, where each switch M1 to M6 may be semiconductor components such as BJTs, MOSFETs, IGBTs or GTOs [1–4]. The switches M1 to M6 are as marked in Figure 4.1. Assume gate drives in Figure 4.1 have a period T, so that the output voltage of this inverter has a frequency of $(1/T)$ Hz. Pulse width modulation (PWM) is *not* considered here for simplicity. Then, each gate drive instance of turn ON and turn OFF duration in degrees are as marked in Figure 4.1. Each gate drive for switches M1 to M6 in order has a pulse ON duration of $(T/2)$ s or 180° and their instance or start of turn ON time between consecutive switches from M1 to M6 differ by $(T/6)$ s or 60°. A three-phase balanced resistive load is used. The DC link voltage V_{dc} is 100 V and the output frequency of this inverter is 50 Hz. The PSIM simulation of the three-phase 180° mode 50 Hz inverter is shown in Figures 4.2a–c and 4.3a–c. Figure 4.2a–c show, respectively, three-phase line-to-ground, line-to-line and line-to-neutral voltages.

57

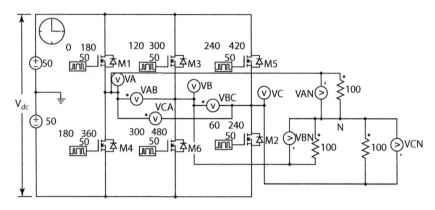

FIGURE 4.1
Three-phase 180° mode inverter.

The harmonic spectrum of the line-to-ground, line-to-line and line-to-neutral voltages are shown in Figure 4.3a–c respectively.

4.2.1 Analysis of Line-to-Line Voltage

Analysis of the line-to-line voltage of the three-phase 180° mode inverter is carried out in this section. The line-to-line voltage V_{ry} is shown in Figure 4.2b. The theoretical derivation of the line-to-neutral voltage waveform is shown in Section 4.2.2. Shifting the axis of Figure 4.2b by $\pi/6$ makes the function $f(-t) = -f(+t)$ odd. The even harmonics are zero. The Fourier coefficient b_n can be defined as follows:

$$b_n = \frac{4}{\pi} * \left[\int_{\pi/6}^{\pi/2} V_{dc} * \sin(n.\omega.t).d(\omega.t) \right] \quad (4.1)$$

$$= \frac{4.V_{dc}}{n.\pi} * \left[\cos\left(\frac{n.\pi}{6}\right) \right] \quad \text{for} \quad n = 1, 3, 5, 7\ldots$$

The line-to-line voltage V_{ry} can be expressed by a Fourier series, as follows:

$$V_{ry} = \sum_{1,3,5,7,\ldots}^{\infty} \frac{4.V_{dc}}{n.\pi} * \left[\cos\left(\frac{n.\pi}{6}\right) * \sin\left(n.\omega.t + \frac{n.\pi}{6}\right) \right] \quad (4.2)$$

The RMS value of V_{ry} is given as follows:

$$V_{ry\_rms} = \sqrt{\frac{1}{\pi} * \int_0^{2\pi/3} V_{dc}^2 * d(\omega.t)}$$

$$= \sqrt{\frac{2}{3}} * V_{dc} = 0.8165 * V_{dc} \quad (4.3)$$

Interactive Models for DC to AC Converters

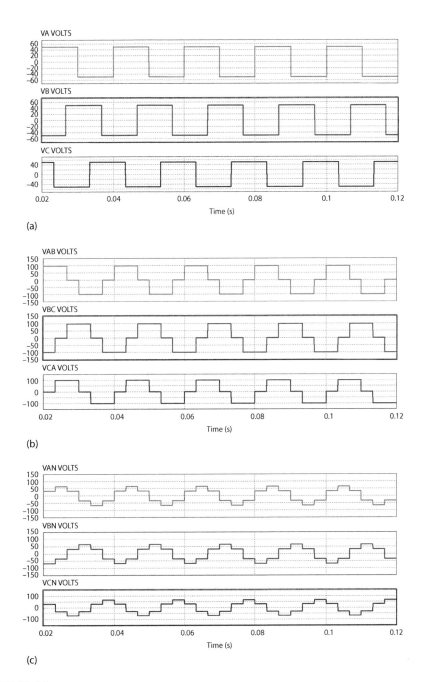

FIGURE 4.2
Three-phase 180° mode inverter. (a) Line-to-ground voltages. (b) Line-to-line voltages. (c) Line-to-neutral voltages.

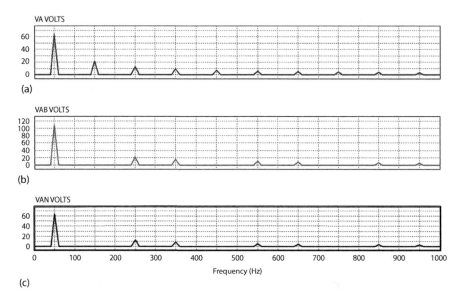

FIGURE 4.3
Three-phase 180° mode inverter: harmonic spectrum of (a) line-to-ground voltage, (b) line-to-line voltage, and (c) line-to-neutral voltage.

The RMS value of the fundamental component of V_{ry} is given as follows:

$$V_{ry1\_rms} = \frac{4.V_{dc}.\cos\left(\frac{\pi}{6}\right)}{\sqrt{2}.\pi} = \frac{\sqrt{6}}{\pi} * V_{dc} \qquad (4.4)$$

The peak value of the fundamental component of the line-to-line voltage of a three-phase 180° mode inverter works out to 110.3 V for a DC link voltage of 100 V, which agrees with the one displayed in Figure 4.3b.

4.2.2 Analysis of Line-to-Neutral Voltage

In the following section, the root mean square (RMS) value and fast Fourier transform (FFT) of the line-to-neutral voltage of a six-step 180° mode inverter are derived.

For clarity, the line-to-neutral voltage of a three-phase 180° mode inverter is derived by simulation using PSIM, shown in Figure 4.1. A resistive star-connected load is used for convenience only, as an inductive load may give rise to voltage spikes in the waveform. The simulated line-to-neutral voltage is shown in Figure 4.2c. A three-phase 50 Hz waveform is generated. The waveform can be theoretically derived as follows:

By taking the ground g as the midpoint of the two DC voltage sources, the line-to-ground voltages V_{rg}, V_{yg} and V_{bg} can be easily derived for three-phase

Interactive Models for DC to AC Converters

$180°$ mode gate drive, by observation of Figure 4.1. This waveform is described as follows:

$$V_{rg} = +\frac{V_{dc}}{2} \quad \text{for} \quad 0 \le \omega t \le \pi$$

$$= -\frac{V_{dc}}{2} \quad \text{for} \quad \pi \le \omega t \le 2\pi \tag{4.5}$$

$$V_{yg} = +\frac{V_{dc}}{2} \quad \text{for} \quad 2\pi/3 \le \omega t \le 5\pi/3$$

$$= -\frac{V_{dc}}{2} \quad \text{for} \quad 0 \le \omega t \le 2\pi/3 \quad \text{and} \tag{4.6}$$

$$\text{for} \quad 5\pi/3 \le \omega t \le 2\pi$$

$$V_{bg} = +\frac{V_{dc}}{2} \quad \text{for} \quad 0 \le \omega t \le \pi/3 \quad \text{and}$$

$$\text{for} \quad 4\pi/3 \le \omega t \le 2\pi \tag{4.7}$$

$$= -\frac{V_{dc}}{2} \quad \text{for} \quad \pi/3 \le \omega t \le 4\pi/3$$

Equations 4.5 to 4.7 confirm well the simulated waveform for three-phase line-to-ground voltage shown in Figure 4.2a.

Then, we have the following line-to-line voltage values:

$$V_{ry} = V_{rg} - V_{yg} \tag{4.8}$$

$$V_{yb} = V_{yg} - V_{bg} \tag{4.9}$$

$$V_{br} = V_{bg} - V_{rg} \tag{4.10}$$

From Equations 4.8 to 4.10, it is possible to derive the three-phase line-to-neutral voltages of a star-connected load, using the following relation:

$$\begin{vmatrix} V_{rn} \\ V_{yn} \\ V_{bn} \end{vmatrix} = \frac{1}{3} * \begin{vmatrix} 1 & 0 & -1 \\ -1 & 1 & 0 \\ 0 & -1 & 1 \end{vmatrix} * \begin{vmatrix} V_{ry} \\ V_{yb} \\ V_{br} \end{vmatrix} \tag{4.11}$$

Such a theoretically derived waveform agrees well with the one shown in Figure 4.2c.

The waveform derived by using Equations 4.5 to 4.11 gives the following coordinates for V_{rn}:

$(0, +V_{dc}/3)$, $(\pi/3, +V_{dc}/3)$, $(\pi/3, +2V_{dc}/3)$, $(2\pi/3, +2V_{dc}/3)$, $(2\pi/3, +V_{dc}/3)$, $(\pi, +V_{dc}/3)$, $(\pi, -V_{dc}/3)$, $(4\pi/3, -V_{dc}/3)$, $(4\pi/3, -2V_{dc}/3)$, $(5\pi/3, -2V_{dc}/3)$, $(5\pi/3, -V_{dc}/3)$, $(2\pi, -V_{dc}/3)$, $(2\pi, +V_{dc}/3)$ and so on. These coordinates agree well with Figure 4.2c. Referring to Figure 4.2c and noting the coordinates given here, we have the following equation for the RMS value of line-to-neutral voltage:

$$V_{rn\_rms} = \sqrt{\frac{1}{\pi} * \left[\frac{V_{dc}^2}{9} * \int_0^{\pi/3} d(\omega t) + \frac{4.V_{dc}^2}{9} * \int_{\pi/3}^{2\pi/3} d(\omega t) + \frac{V_{dc}^2}{9} * \int_{2\pi/3}^{\pi} d(\omega t) \right]} \qquad (4.12)$$

$$= \frac{\sqrt{2} * V_{dc}}{3} \qquad (4.13)$$

The Fourier transform of the line-to-neutral voltage V_{rn} is derived as follows:

Referring to Figure 4.2c, the axis of symmetry is the origin where $f(-t) = -f(+t)$ and the function is odd. Thus, the Fourier coefficient b_n is given by the following:

$$b_n = \frac{2}{\pi} * \left[\int_0^{\pi/3} \frac{V_{dc}}{3} * \sin(n.\omega t).d(\omega t) + \int_{\pi/3}^{2\pi/3} \frac{2V_{dc}}{3} * \sin(n.\omega t).d(\omega t) \right.$$

$$\left. + \int_{2\pi/3}^{\pi} \frac{V_{dc}}{3} * \sin(n.\omega t).d(\omega t) \right] \qquad (4.14)$$

$$= \frac{2V_{dc}}{3\pi.n} * \left[1 - \cos\left(\frac{n.\pi}{3}\right) + 2 * \left(\cos\left(\frac{n.\pi}{3}\right) - \cos\left(\frac{2n.\pi}{3}\right) \right) \right.$$

$$\left. + \cos\left(\frac{2n.\pi}{3}\right) - \cos(n.\pi) \right]$$

$$= \sum_{n=1,5,7\ldots}^{\infty} \frac{2V_{dc}}{3\pi.n} * \left[1 + \cos\left(\frac{n.\pi}{3}\right) - \cos\left(\frac{2n.\pi}{3}\right) - \cos(n.\pi) \right] \qquad (4.15)$$

Substituting $n = 1,2,3,\ldots$, we have the following general expression for the FFT of the line-to-neutral voltage:

$$V_{rn} = \frac{2V_{dc}}{\pi} * \left[\sin(\omega t) + \frac{1}{5} * \sin(5\omega t) + \frac{1}{7} * \sin(7\omega t) + \ldots \right] \qquad (4.16)$$

Interactive Models for DC to AC Converters 63

The RMS value of the fundamental component of V_{rn} is given here:

$$V_{rn1(\text{rms})} = \frac{\sqrt{2}.V_{dc}}{\pi} \tag{4.17}$$

For a DC link voltage of 100 V, the peak fundamental component of V_{rn} given by Equation 4.17 is 63.675 V, which agrees well with the simulation results shown in Figure 4.3c.

4.2.3 Total Harmonic Distortion

The output voltage of the three-phase inverter contains harmonics and, hence, the waveform is distorted. Total harmonic distortion (THD) is a measure of the closeness between a distorted waveform and its fundamental component. Thus, THD is the ratio of the RMS value of the distorted waveform and the RMS value of its fundamental component. The THD of the voltage waveform is defined here:

$$\text{THD} = \sqrt{\left(\frac{V_{S\_RMS}}{V_{S1\_RMS}}\right)^2 - 1} \tag{4.18}$$

The same definition for THD given by Equation 4.18 holds good for currents as well.

Using Equations 4.3, 4.4, 4.13, 4.17 and 4.18, the THD of the line-to-line and line-to-neutral voltages are calculated as follows:

$$\text{THD of } V_{LL} = \sqrt{\left(\frac{2.V_{dc}^2 / 3}{6V_{dc}^2 / \pi^2}\right) - 1} = 30.9\% \tag{4.19}$$

$$\text{THD of } V_{LN} = \sqrt{\left(\frac{2.V_{dc}^2 / 9}{2.V_{dc}^2 / \pi^2}\right) - 1} = 30.9\% \tag{4.20}$$

4.2.4 Model for Three-Phase 180° Mode Inverter

The model of the three-phase 180° mode inverter is shown in Figure 4.4 [5–8]. The various dialogue boxes where users may enter data are given in Figure 4.5. The dialogue boxes mainly correspond to entering the frequency of switching the inverter and its DC link voltage. The phase advance is entered as zero. The details of the various subsystems of Figure 4.4 are shown in Figure 4.6a to d. The functions of the various subsystems are explained here:

Figure 4.6a is the subsystem for the three-phase 180° mode inverter gate drive parameters. The corresponding dialogue box is shown in Figure 4.5.

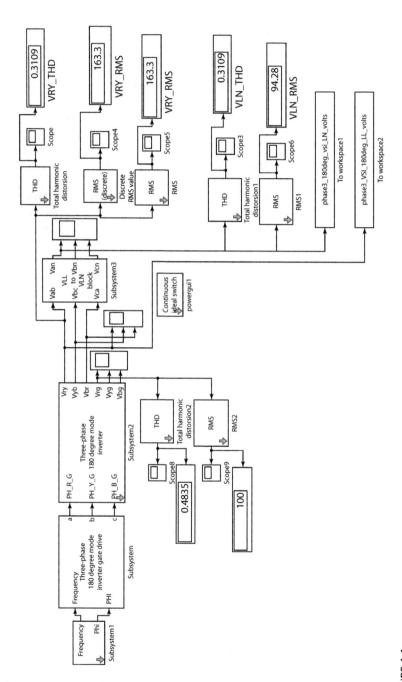

FIGURE 4.4
Model of a three-phase 180° mode inverter.

Interactive Models for DC to AC Converters

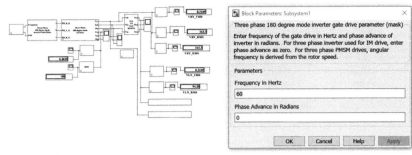

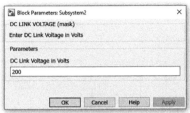

FIGURE 4.5
Model of a three-phase 180° mode inverter with interactive blocks.

The frequency entered is 60 Hz and the phase advance is zero. This frequency is internally multiplied by 2π to obtain the angular frequency to generate the sine wave and the gate drive for the inverter.

The outputs of Figure 4.6a are passed to the four-input mux in Figure 4.6b. In Figure 4.6b, the three Fcn blocks, Fcn, Fcn1 and Fcn2, connected to the four-input mux, generate a three-phase sine wave at the frequency entered with the phase advance added. The three-phase sine-wave output is connected to three Fcn blocks, Fcn3, Fcn4 and Fcn5, which compare the respective $u(1)$ inputs with 0, generate logic 1 if $u(1)$ is greater than or equal to zero, else output logic 0. The output of the three Fcn blocks Fcn3, Fcn4 and Fcn5 are the switch functions a, b and c.

The output a, b and c of the three Fcn blocks in Figure 4.6b are passed to the second $u(2)$ input of the three threshold switches, Switch, Switch1 and Switch2, respectively, in Figure 4.6c. One half of the DC link voltages $+V_{dc}/2$ and $-V_{dc}/2$ is generated using a constant block and two gain blocks, Gain and Gain1, as shown in Figure 4.6c. The $u(1)$ and $u(3)$ input to these three threshold switches are $+V_{dc}/2$ and $-V_{dc}/2$, respectively. When the $u(2)$ input to these three threshold switches is greater than or equal to the threshold value of 0.5, the outputs of these switches equal $+V_{dc}/2$, else the outputs equal $-V_{dc}/2$. Thus, the three switches generate the three-phase line-to-ground voltages V_{rg}, V_{yg} and V_{bg}, respectively. These line-to-ground voltages are subtracted in pairs using subtract blocks Sum2, Sum3 and Sum4 to generate the three-phase line-to-line voltages V_{ry}, V_{yb} and V_{br}, respectively. Figure 4.6c opens up the dialogue box shown in Figure 4.5 for the DC link voltage. The DC link voltage entered is 200 V.

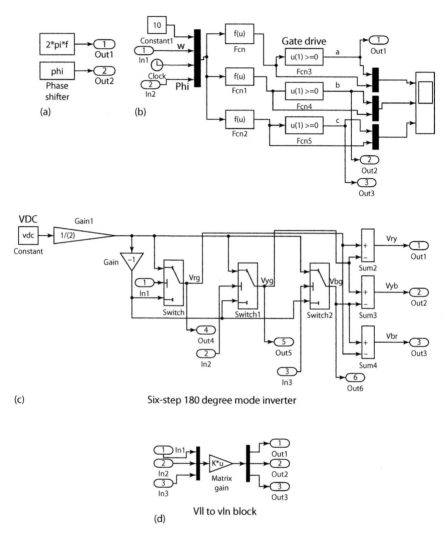

FIGURE 4.6
(a–d) Three-phase 180° mode inverter model subsystems.

The three-phase line-to-line voltages V_{ry}, V_{yb} and V_{br} are respectively applied to the three-input mux shown in Figure 4.6d. The Matrix Gain block in Figure 4.6d multiplies V_{ry}, V_{yb} and V_{br} by the matrix defined in Equation 4.11 to obtain the line-to-neutral voltages V_{rn}, V_{yn} and V_{bn} as the output of the demux block.

The THD and RMS values of the line-to-line and line-to-neutral voltages are measured using appropriate measurement blocks from the Power Systems block set. The Powergui from the Power Systems block set is used to display the harmonic spectrum of the line-to-neutral and line-to-line voltages.

Interactive Models for DC to AC Converters

4.2.5 Simulation Results

The simulation of the three-phase 180° mode inverter model was carried out using the ode23t (mod. stiff/trapezoidal) solver. The three-phase line-to-ground voltages, line-to-line voltages and line-to-neutral voltages are shown in Figures 4.7a to c, 4.8a to c and 4.9a to c, respectively. The RMS and THD values of the line-to-line voltages are shown in Figure 4.10a and b and those corresponding to the line-to-neutral voltages are shown in Figure 4.11a and b, respectively. The harmonic spectrum of the line-to-line voltages is shown in Figure 4.12a and b and that corresponding to the line-to-neutral voltages is shown in Figure 4.13a and b, respectively. The results observed for the THD and RMS values of the line-to-line and line-to-neutral voltages by simulation

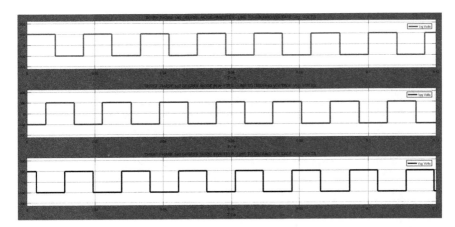

FIGURE 4.7
Three-phase 180° mode inverter. Line-to-ground voltage: (a) V_{rg} V, (b) V_{yg} V and (c) V_{bg} V.

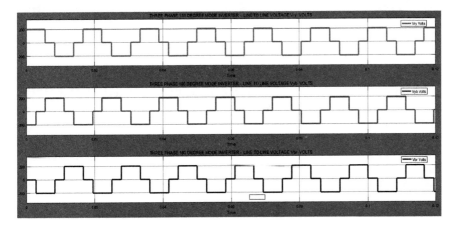

FIGURE 4.8
Three-phase 180° mode. Line-to-line voltage: (a) V_{ry} V, (b) V_{yb} V and (c) V_{br} V.

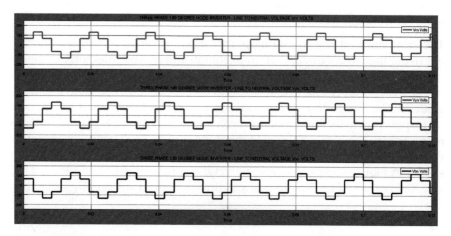

FIGURE 4.9
Three-phase 180° mode inverter. Line-to-neutral voltage: (a) V_{rn} V, (b) V_{yn} V and (c) V_{bn} V.

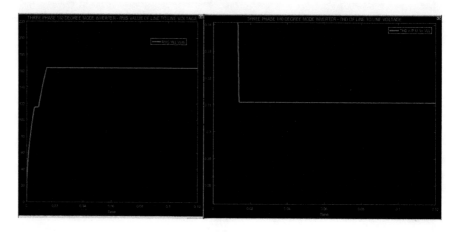

FIGURE 4.10
Three-phase 180° mode inverter. (a) RMS value of line-to-line voltage. (b) THD of line-to-line voltage.

are also tabulated in Table 4.1. The theoretically calculated THD and RMS values of the line-to-line and line-to-neutral voltages using Equations 4.3, 4.4, 4.13 and 4.17 are tabulated in Table 4.2.

4.3 Three-Phase 120° Mode Inverter

The three-phase inverter shown in Figure 4.1 can also be driven by 120° mode gate drive.

Interactive Models for DC to AC Converters 69

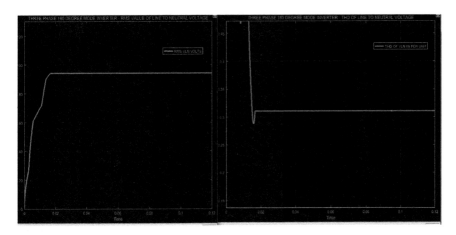

FIGURE 4.11
Three-phase 180° mode inverter. (a) RMS value of line-to-neutral voltage. (b) THD of line-to-neutral voltage.

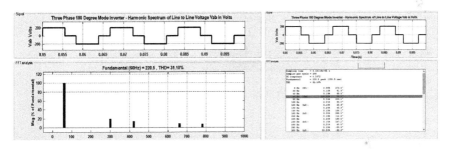

FIGURE 4.12
(a,b) Three-phase 180° mode inverter: harmonic spectrum of line-to-line voltage V_{ab} in volts.

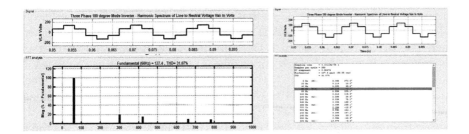

FIGURE 4.13
(a,b) Three-phase 180° mode inverter: harmonic spectrum of line-to-neutral voltage V_{an} in volts.

TABLE 4.1

Three-Phase 180° Mode Inverter: Simulation Results

Sl.No.	Frequency(Hz)	DC Link Voltage (V)	THD of V_{LL} (Per Unit)	RMS V_{LL} (V)	THD of V_{LN} (Per Unit)	RMS V_{LN} (V)
1	60	200	0.3109	163.3	0.3109	94.28

TABLE 4.2

Three-Phase 180° Mode Inverter: Calculated Values

Sl. No.	Frequency (Hz)	DC Link Voltage (V)	THD of V_{LL} (Per Unit)	RMS V_{LL} (V)	THD of V_{LN} (Per Unit)	RMS V_{LN} (V)
1	60	200	0.309	163.28	0.309	94.27

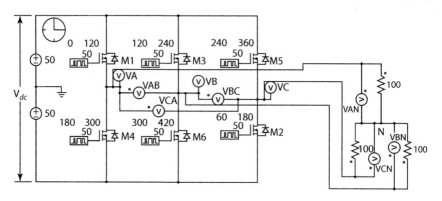

FIGURE 4.14
Three-phase inverter: 120° mode.

The three-phase 120° mode inverter topology is shown in Figure 4.14. The switches M1 to M6 are as marked in Figure 4.14. Assume that the gate drives in Figure 4.14 have a period T, so that the output voltage of this inverter has a frequency of $(1/T)$ Hz. PWM is not considered here for simplicity. Then, each gate drive instance of turn ON and turn OFF duration in degrees are as marked in Figure 4.14. Each gate drive M1 to M6 in order has a pulse ON duration of $(T/3)$ s or 120° and their instance or start of turn ON time between successive switches from M1 to M6 differ by $(T/6)$ s or 60°. A three-phase balanced resistive load is used. The DC link voltage V_{dc} is 100 V and the output frequency of this inverter is 50 Hz. The PSIM simulation of the three-phase 120° mode 50 Hz inverter is shown in Figures 4.15a to c, 4.16a to c and 4.17a to c. Figure 4.18a to c shows the harmonic spectrum of the three-phase line-to-ground, line-to-line and line-to-neutral voltages, respectively.

Interactive Models for DC to AC Converters 71

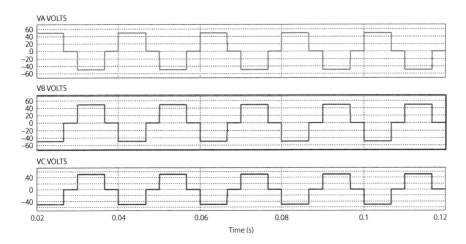

FIGURE 4.15
(a–c) Three-phase 120° mode inverter: line-to-ground voltage.

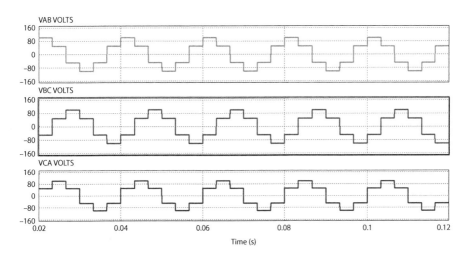

FIGURE 4.16
(a–c) Three-phase 120° mode inverter: line-to-line voltage.

4.3.1 Analysis of Line-to-Line Voltage

Analysis of the line-to-line voltage of the three-phase 120° mode inverter is carried out in this section. The line-to-line voltage V_{ry} is shown in Figure 4.16a. The theoretical derivation of the line-to-neutral voltage waveform is shown in Section 4.3.2. Shifting the axis of Figure 4.16a by $\pi/3$ makes the

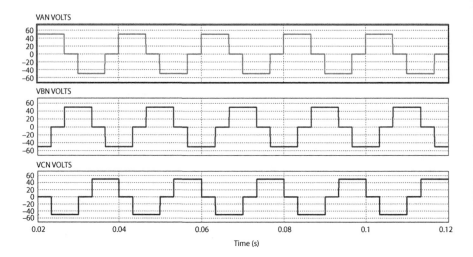

FIGURE 4.17
(a–c) Three-phase 120° mode inverter: line-to-neutral voltage.

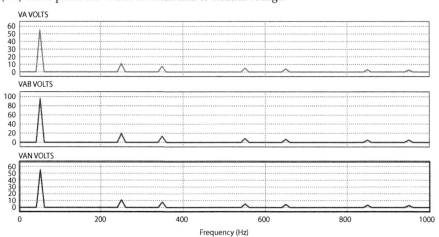

FIGURE 4.18
Three-phase 120° mode inverter: harmonic spectrum of (a) line-to-ground voltage, (b) line-to-line voltage and (c) line-to-neutral voltage.

function $f(-t) = -f(+t)$ odd. The even harmonics are zero. The Fourier coefficient b_n can be defined as follows:

$$b_n = \frac{4}{\pi} * \left[\int_0^{\pi/3} \frac{V_{dc}}{2} * \sin(n\omega t).d(\omega t) + \int_{\pi/3}^{\pi/2} V_{dc} * \sin(n\omega t).d(\omega t) \right]$$

$$= \sum_{n=1,3,5,7,\ldots}^{\infty} \frac{2V_{dc}}{\pi.n} * \left[1 + \cos\left(\frac{n.\pi}{3}\right) \right]$$

(4.21)

Interactive Models for DC to AC Converters 73

The line-to-line voltage V_{ry} can be expressed by a Fourier series, as follows:

$$V_{ry} = \sum_{n=1,3,5,7,\ldots}^{\infty} \frac{2V_{dc}}{\pi.n} * \left[1 + \cos\left(\frac{n.\pi}{3}\right) * \sin\left(n.\omega t + \frac{n.\pi}{3}\right)\right] \quad (4.22)$$

The RMS value of V_{ry} is given as follows:

$$V_{ry\_rms} = \sqrt{\frac{1}{\pi} * \left[\int_{0}^{\pi/3} V_{dc}^2 * d(\omega t) + \int_{\pi/3}^{2\pi/3} \frac{V_{dc}^2}{4} * d(\omega t) + \int_{2\pi/3}^{\pi} \frac{V_{dc}^2}{4} * d(\omega t)\right]}$$

$$= \frac{V_{dc}}{\sqrt{2}} \quad (4.23)$$

The RMS value of the fundamental component of V_{ry} from Equation 4.22 is given here:

$$V_{ry1\_rms} = \frac{3}{\sqrt{2}} * \left[\frac{V_{dc}}{\pi}\right] \quad (4.24)$$

The peak value of the fundamental component of V_{ry} shown in Figure 4.18b agrees well with Equation 4.24 for a DC link voltage of 100 V.

4.3.2 Analysis of Line-to-Neutral Voltage

In the following section, the RMS value and FFT of the line-to-neutral voltage of a six-step 120° mode inverter are derived. For clarity, the line-to-neutral voltage of a three-phase 120° mode inverter is derived by simulation using PSIM, shown in Figure 4.14. A resistive star-connected load is used for convenience only, as an inductive load may give rise to voltage spikes in the waveform. The simulated line-to-neutral voltage is shown in Figure 4.17a to c. A three-phase 50 Hz waveform is generated. The DC link voltage used is 100 V. The waveform can be theoretically derived as follows:

Referring to Figure 4.14, by taking the ground g as the midpoint of the two DC voltage sources forming the DC link, the line-to-ground voltages V_{rg}, V_{yg} and V_{bg} can be easily derived for the three-phase 120° mode gate drive, by observation of Figure 4.14 and its gate drive pattern discussed here. This waveform is described as follows:

$$V_{rg} = +V_{dc}/2 \quad \text{for} \quad 0 \le \omega t \le 2\pi/3$$

$$= 0 \quad \text{for} \quad 2\pi/3 \le \omega t \le \pi$$

$$= -V_{dc}/2 \quad \text{for} \quad \pi \le \omega t \le 5\pi/3 \quad (4.25)$$

$$= 0 \quad \text{for} \quad 5\pi/3 \le \omega t \le 2\pi$$

$$V_{yg} = +V_{dc}/2 \quad \text{for} \quad 2\pi/3 \leq \omega t \leq 4\pi/3$$
$$= 0 \quad \text{for} \quad 4\pi/3 \leq \omega t \leq 5\pi/3$$
$$= -V_{dc}/2 \quad \text{for} \quad 5\pi/3 \leq \omega t \leq 2\pi \quad \text{and} \qquad (4.26)$$
$$\text{for} \quad 0 \leq \omega t \leq \pi/3$$
$$= 0 \quad \text{for} \quad \pi/3 \leq \omega t \leq 2\pi/3$$
$$V_{bg} = +V_{dc}/2 \quad \text{for} \quad 4\pi/3 \leq \omega t \leq 2\pi$$
$$= 0 \quad \text{for} \quad 0 \leq \omega t \leq \pi/3$$
$$= -V_{dc}/2 \quad \text{for} \quad \pi/3 \leq \omega t \leq \pi \qquad (4.27)$$
$$= 0 \quad \text{for} \quad \pi \leq \omega t \leq 4\pi/3$$

From Equations 4.25, 4.26 and 4.27, the line-to-line voltages can be expressed as follows:

$$V_{ry} = V_{rg} - V_{yg} \qquad (4.28)$$

$$V_{yb} = V_{yg} - V_{bg} \qquad (4.29)$$

$$V_{br} = V_{bg} - V_{rg} \qquad (4.30)$$

The line-to-neutral voltages V_{rn}, V_{yn} and V_{bn} can be derived as follows:

$$\begin{vmatrix} V_{rn} \\ V_{yn} \\ V_{bn} \end{vmatrix} = \frac{1}{3} * \begin{vmatrix} 1 & 0 & -1 \\ -1 & 1 & 0 \\ 0 & -1 & 1 \end{vmatrix} * \begin{vmatrix} V_{ry} \\ V_{yb} \\ V_{br} \end{vmatrix} \qquad (4.31)$$

The waveform derived using Equation 4.31 looks much the same as shown in Figure 4.17a to c.

The waveform derived by using Equations 4.25 to 4.31 gives the following coordinates for V_{rn}:

(0, 0), (0, $+V_{dc}/2$), ($2\pi/3$, $+V_{dc}/2$), ($2\pi/3$, 0), (π, 0), (π, $-V_{dc}/2$), ($5\pi/3$, $-V_{dc}/2$), ($5\pi/3$, 0), (2π, 0), (2π, $+V_{dc}/2$) and so on.

Referring to Figure 4.17a and noting the coordinates given here, we have the following equation for the RMS value of the line-to-neutral voltage:

$$V_{rn\_rms} = \sqrt{\frac{1}{\pi} * \int_0^{2\pi/3} \left[\frac{V_{dc}^2}{4}\right].d(\omega t)} = \frac{V_{dc}}{\sqrt{6}} \qquad (4.32)$$

Interactive Models for DC to AC Converters   75

The Fourier transform of the line-to-neutral voltage V_{rn} is derived as follows:

Referring to Figure 4.17a, the axis of symmetry is shifted by $\pi/6$ from the origin where $f(-t)=-f(+t)$ and the function is odd. The even harmonics are zero. Thus, the Fourier coefficient b_n is given by the following equation:

$$b_n = \frac{2}{\pi} * \left[\int_{\pi/6}^{5\pi/6} \frac{V_{dc}}{2} * \sin(n.\omega t).d(\omega t) \right]$$

$$= \frac{2V_{dc}}{n.\pi} * \cos\left(\frac{n.\pi}{6}\right) \text{ for } n = 1,3,5,7\ldots$$

(4.33)

The line-to-neutral voltage V_{rn} can be expressed by a Fourier series, as follows:

$$V_{rn} = \sum_{1,3,5,7,\ldots}^{\infty} \frac{2V_{dc}}{n.\pi} * \cos\left(\frac{n.\pi}{6}\right) * \sin\left(n.\omega t + \frac{n.\pi}{6}\right)$$

(4.34)

The RMS value of the fundamental component of V_{rn} from Equation 4.34 is given here:

$$V_{rn1\_rms} = \sqrt{\frac{3}{2}} * \left[\frac{V_{dc}}{\pi}\right]$$

(4.35)

The peak value of the fundamental component of V_{rn} shown in Figure 4.18c agrees well with Equation 4.35.

4.3.3 Total Harmonic Distortion

The THD of the voltage wave form is defined by Equation 4.18 in Section 4.2.3.

Using Equations 4.23, 4.24, 4.32 and 4.35, the THD of the line-to-line and line-to-neutral voltages are calculated as follows:

$$\text{THD of } V_{LL} = \sqrt{\frac{V_{dc}^2/2}{9.V_{dc}^2/2.\pi^2} - 1} = 30.9\%$$

(4.36)

$$\text{THD of } V_{LN} = \sqrt{\frac{V_{dc}^2/6}{3.V_{dc}^2/2.\pi^2} - 1} = 30.9\%$$

(4.37)

76 *Power Electronic Converters*

4.3.4 Model of Three-Phase 120° Mode Inverter

The model of the three-phase 120° mode inverter is shown in Figure 4.19 [8]. The various dialogue boxes where the user may enter data are given in Figure 4.20. The dialogue boxes mainly correspond to entering the frequency of switching the inverter and its DC link voltage. The phase advance is entered as zero. The details of the various subsystems of Figure 4.19 are shown in Figure 4.21a to d. The functions of the various subsystems are explained next.

Figure 4.21a corresponds to the dialogue box with the name 'Three-phase 120 degree mode inverter gate drive' shown in Figure 4.20. A value of 60 is entered in the dialogue box corresponding to the inverter switching frequency and zero corresponding to the phase advance. This frequency is internally multiplied by 2π to obtain the angular switching frequency of the inverter. PWM is not considered in this model.

The three-phase 120° mode inverter gate drive block is shown in Figure 4.21b. The four inputs to the mux are the arbitrary constant, angular switching frequency, time and switching angle advance. The three-phase sine-wave AC with angular frequency corresponding to the frequency entered with phase advance added is generated using the Fcn1, Fcn2 and Fcn3 blocks. Each of the three-phase AC outputs are then compared in Schmitt trigger relay comparators; Relay1, 2 and 3 are used as zero-crossing comparators. The output of Relay1, 2 and 3 becomes logic 1 when the respective input crosses zero and becomes positive and their output becomes logic 0 when their respective input crosses zero and becomes negative.

The three-phase 120° mode inverter block is shown in Figure 4.21c. The a, b, c outputs of Relay1, 2 and 3 are passed in pairs to three Fcn blocks, Fcn6, Fcn7 and Fcn8, using three two-input mux. The three Fcn blocks Fcn6, 7 and 8 subtract the two inputs to their respective mux. Each of the three Subtract blocks are passed to a pair of zero-comparison blocks. The Fcn3 block outputs logic 1 when the input $u(1)$ is greater than zero, else outputs logic 0. Similarly, the Fcn4 block outputs logic 1 when the input $u(1)$ is less than zero, else outputs logic 0. The same logic holds good for zero-comparison block pairs Fcn5–Fcn9 and Fcn10–Fcn11. The outputs of Fcn3 and Fcn4 are passed to the $u(2)$ input of threshold Switch and Switch1, respectively. The $u(1)$ inputs for Switch and Switch1 are, respectively, $+V_{dc}/2$ and $-V_{dc}/2$ and the $u(3)$ inputs to both switches are zeros. Both Switch and Switch1 output $u(1)$ when $u(2)$ is greater than or equal to the threshold value of 0.5, else their output is $u(3)$. The output of Switch and Switch1 are connected to the Add block. The same principle holds good for the switch pairs corresponding to the Fcn5–Fcn9 and Fcn10–Fcn11 block pairs. Thus, each of the Add blocks generate the line-to-ground voltages, V_{rg}, V_{yg} and V_{bg}, respectively. The three-phase line-to-ground voltages are subtracted in pairs using the three Subtract blocks marked Add3, Add4 and Add5 to generate the three-phase line-to-line voltages V_{ry}, V_{yb} and V_{br}.

Interactive Models for DC to AC Converters 77

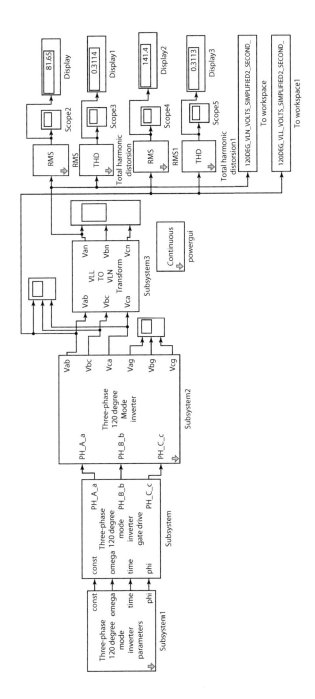

FIGURE 4.19
Three-phase 120° mode inverter.

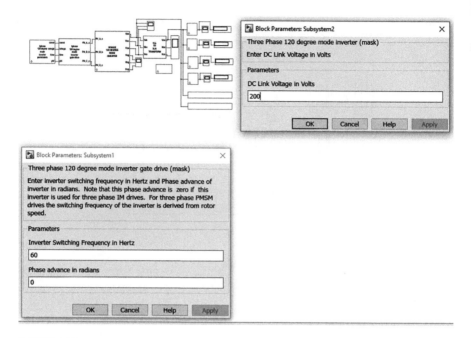

FIGURE 4.20
Three-phase 120° mode inverter with interactive blocks.

The line-to-line to line-to-neutral voltage transformation block is shown in Figure 4.21d. The three line-to-line voltages V_{ry}, V_{yb} and V_{br} are passed to a Matrix Gain block using a three-input mux. The gain matrix K of the Matrix Gain block is defined by Equation 4.31. The three-phase line-to-neutral voltages V_{rn}, V_{yn} and V_{bn} are obtained at the output of the demux block.

Apart from this, the RMS and THD blocks from the Power Systems block set are used to measure the RMS value and THD of the line-to-line and line-to-ground voltages. Scopes are used to display the three-phase line-to-ground, line-to-line and line-to-neutral voltages. The Powergui from the Power Systems block set is used for harmonic analysis of the line-to-line and line-to-neutral voltages.

4.3.5 Simulation Results

A simulation of the model of the three-phase 120° mode inverter was carried out using the ode23t (mod. stiff/trapezoidal) solver. The DC link voltage and frequency entered were 200 V and 60 Hz, respectively. The three-phase line-to-ground voltages, line-to-line voltages and line-to-neutral voltages are shown in Figure 4.22a to c, Figures 4.23a to c and 4.24a to c, respectively. The RMS and THD values of the line-to-line voltages are shown in Figure 4.25a and b and those corresponding to the line-to-neutral voltages are shown in Figure 4.26a and b, respectively. The harmonic spectrum of the line-to-line

Interactive Models for DC to AC Converters

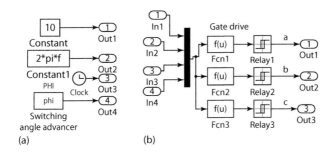

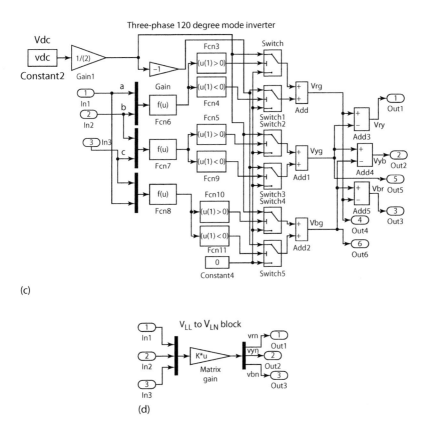

FIGURE 4.21
(a–d) Three-phase 120° mode inverter subsystems.

voltages is shown in Figure 4.27a and b and that corresponding to the line-to-neutral voltages is shown in Figure 4.28a and b, respectively. These simulation results are also tabulated in Table 4.3. The theoretically calculated values for the THD and RMS values of the line-to-line and line-to-neutral voltages using Equations 4.23, 4.24, 4.32, 4.35, 4.36 and 4.37 are tabulated in Table 4.4.

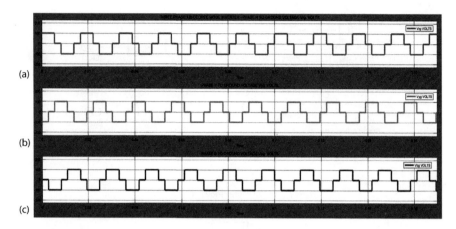

FIGURE 4.22
Three-phase 120° mode inverter. (a) Phase R to ground voltage V_{rg} V. (b) Phase Y to ground voltage V_{yg} V. (c) Phase B to ground voltage V_{bg} V.

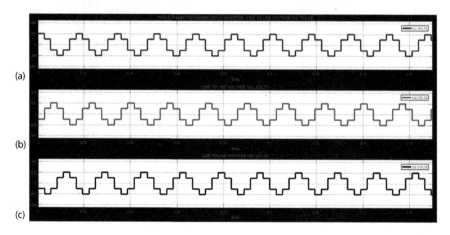

FIGURE 4.23
Three-phase 120° mode inverter. (a) Line-to-line voltage V_{ry} V. (b) Line-to-line voltage V_{yb} V. (c) Line-to-line voltage V_{br} V.

4.4 Three-Phase Sine PWM Technique

Many industrial applications require voltage control of inverters to maintain constant output voltage irrespective of variations in the input DC voltage and for constant voltage and frequency control requirements. This is achieved by PWM, commonly known as *pulse width modulation (PWM) technique* [1–4].

The sine PWM is the most fundamental and widely used technique for voltage control of inverters. The principle of generation of the gating pulse for a three-phase sine PWM inverter using op-amps is shown in Figure 4.29a,

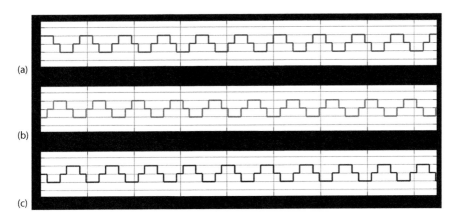

FIGURE 4.24
(a–c) Three-phase 120° mode inverter – line to neutral voltage in volts.

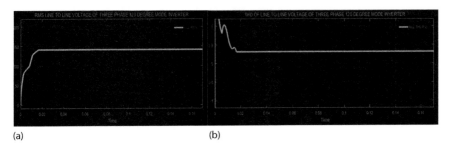

FIGURE 4.25
(a) RMS and (b) THD of line-to-line voltage of three-phase 120° mode inverter.

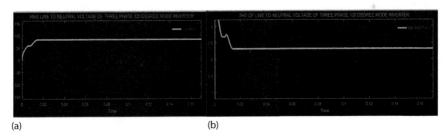

FIGURE 4.26
(a) RMS and (b) THD of line-to-neutral voltage of three-phase 120° mode inverter.

using Microcap 11. Three-phase AC modulating signals at the switching frequency of f_m Hz of the inverter is generated. This three-phase modulating signal is compared with the triangular carrier wave of frequency f_c Hz in a comparator, and the resulting pulse width modulated output pulse and its respective inverted pulse are used to drive the semiconductor switches in each phase of the three-phase inverter. The three-phase 50 Hz AC sine modulating signal, triangular carrier signal of frequency 2 kHz and the resulting

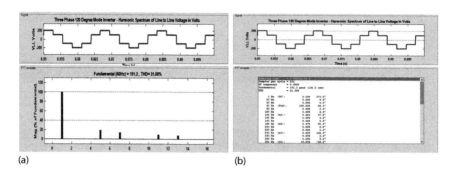

FIGURE 4.27
(a,b) Three-phase 120° mode inverter: harmonic spectrum of line-to-line voltage in volts.

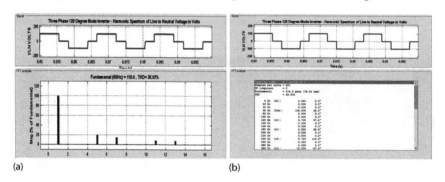

FIGURE 4.28
(a,b) Three-phase 120° mode inverter: harmonic spectrum of line-to-neutral voltage in volts.

TABLE 4.3

Three-Phase 120° Mode Inverter: Simulation Results

Sl. No.	Frequency (Hz)	DC Link Voltage (V)	THD of V_{LL} (Per Unit)	RMS V_{LL} (V)	THD of V_{LN} (Per Unit)	RMS V_{LN} (V)
1	60	200	0.3113	141.4	0.3114	81.65

TABLE 4.4

Three-Phase 120° Mode Inverter: Calculated Values

Sl. No.	Frequency (Hz)	DC Link Voltage (V)	THD of V_{LL} (Per Unit)	RMS V_{LL} (V)	THD of V_{LN} (Per Unit)	RMS V_{LN} (V)
1	60	200	0.309	141.44	0.309	81.66

gate pulse drive for the three-phase inverter are shown in Figure 4.29b by simulation using Microcap 11. The three-phase inverter frequency output will be 50 Hz. If A_m and A_C are the peak values of the sine modulating signal and the triangle carrier and their respective frequencies are f_m and f_C,

Interactive Models for DC to AC Converters

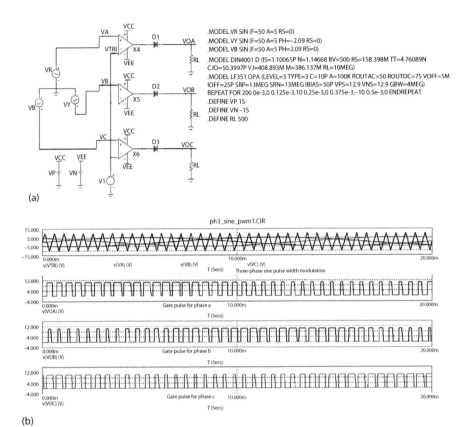

FIGURE 4.29
(a) Three-phase sine PWM and (b) gate pulse for three-phase sine PWM inverter.

then amplitude modulation and frequency modulation indices M and FM are defined as shown here:

$$M = \frac{A_m}{A_C} \quad (4.38)$$

$$FM = \frac{f_C}{f_m} \quad (4.39)$$

The region where $0 < M \leq 1$ is called the *linear region* or *undermodulation region*, and the region where $M > 1$ is called the *overmodulation region*. In the linear region, the maximum value of the fundamental component of the output line-to-line voltage v_{ab1} can be expressed as given in Equation 4.40:

$$v_{ab1} = \frac{M.\sqrt{3}.V_{dc}}{2} \quad \text{for} \quad M \leq 1 \tag{4.40}$$

where V_{dc} is the DC link input voltage of the three-phase inverter.

In the overmodulation region, the maximum value of the nth harmonic component of the line-to-line voltage output of the three-phase inverter is given by Equation 4.41:

$$v_{abn} = \frac{4.\sqrt{3}.V_{dc}}{2.n.\pi} \quad \text{for} \quad M > 1 \tag{4.41}$$

Thus, by varying the value of M, the output voltage of the three-phase inverter can be controlled.

4.4.1 Model for Three-Phase Sine PWM Inverter

The model for the three-phase sine PWM inverter with interactive blocks is shown in Figure 4.30 [7,8]. The various dialogue boxes are shown in Figure 4.30. The various subsystem detailed schematics are shown in Figure 4.31a to e. The various subsystems are explained as follows:

The three subsystems 'Switching Function Generator', 'Three-Phase Sine PWM Generator' and 'Three-Phase VSI' in Figure 4.31a correspond to the dialogue box 'Three-Phase Inverter and Amplitude and Frequency Modulator Unit' and its model subsystems shown in Figure 4.31b to d. In Figure 4.31b, the three-phase AC with amplitude 1 V and frequency f_m (60 Hz) of the sine-wave

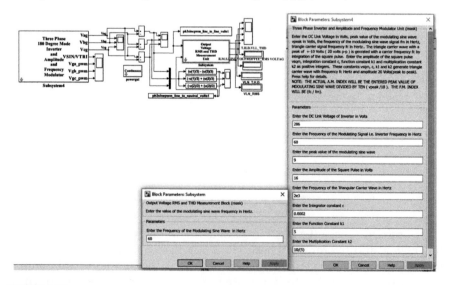

FIGURE 4.30
Model of three-phase sine PWM inverter with interactive blocks.

Interactive Models for DC to AC Converters 85

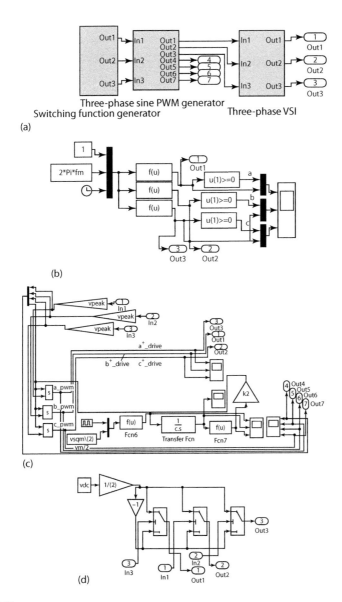

FIGURE 4.31
(a–e) Three-phase sine PWM inverter model subsystems.

modulating signal is generated using three Fcn blocks connected to the mux. The three-phase AC output of this sine-wave modulating signal is passed to three Gain blocks with gain v_{peak} in Figure 4.31c. The output of these three Gain blocks are three-phase sine-wave modulating signals with a peak value of v_{peak}. The triangle carrier wave of frequency f_C (2 kHz) is generated by integration of the square pulse whose amplitude is V_{sqm} and frequency f_C Hz. The Fcn6 block

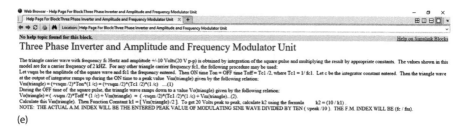

(e)

FIGURE 4.31 (CONTINUED)
(a–e) Three-phase sine PWM inverter model subsystems.

subtracts this square pulse from $V_{sqm}/2$ generating square wave with positive and negative peaks of $+V_{sqm}/2$ and $-V_{sqm}/2$, respectively. This is then multiplied by a constant of integration $1/c$ to obtain a triangle carrier wave with a maximum value of $+V_m$(triangle) and minimum value of zero. This triangle carrier wave is then subtracted from the constant $k1$ to obtain a triangle carrier wave with positive and negative peaks of $+V_m$(triangle)/2 and $-V_m$(triangle)/2, respectively, using the Fcn7 block. The value of $k1$ is $+V_m$(triangle)/2. The output of the Fcn7 block is then multiplied by $k2$ to obtain a triangle carrier wave with positive and negative peaks of $+10$ V and -10 V, respectively. The value of $k2$ is $10/k1$. The values shown in the dialogue box are to generate a carrier wave of 2 kHz frequency. For any other frequency of the carrier wave, the method shown in Figure 4.31e can be used. The values relating to the DC link voltage of the three-phase inverter, modulating signal frequency (i.e. inverter frequency), amplitude of the modulating signal v_{peak}, V_{sqm}, f_C, c, $k1$ and $k2$ are entered in the appropriate box. The amplitude modulation index M is the value entered of the sine modulating signal peak value, $v_{peak}/10$. This triangle carrier wave and the three-phase modulating sine-wave AC are compared in the three relational operator blocks used as comparators. The three comparators output the three-phase gating pulses a_pwm, b_pwm and c_pwm for the three-phase inverter. Figure 4.31d shows the model of the three-phase inverter. The gating pulses a_pwm, b_pwm and c_pwm are passed to the $u(2)$ input of three threshold switches. The $u(1)$ and $u(3)$ inputs to these three switches are the $+V_{dc}/2$ and $-V_{dc}/2$ inputs. The threshold value for these three switches is 0.5. When the respective $u(2)$ input of these three switches becomes logic 1, the output is $+V_{dc}/2$, and when it is logic 0, the output is $-V_{dc}/2$. Thus, three-phase line-to-ground voltage is generated.

4.4.2 Simulation Results

A system model simulation of the three-phase sine PWM inverter was carried out for an amplitude modulation index M of 0.9 in the undermodulation region and for an index M of 1.1 in the overmodulation region. The DC link voltage was 286 V. The sine modulating signal frequency was 60 Hz and the triangle carrier frequency was 2 kHz. The simulation results for $M = 0.9$ are shown in Figure 4.32 and the harmonic spectrum of the line-to-line and line-to-neutral voltages in

Interactive Models for DC to AC Converters

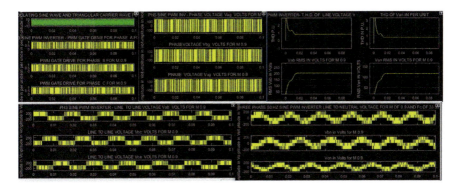

FIGURE 4.32
Three-phase sine PWM inverter simulation results for $M = 0.9$.

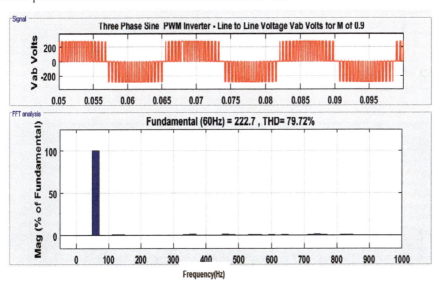

FIGURE 4.33
Three-phase sine PWM inverter: harmonic spectrum of line-to-line voltage for $M = 0.9$.

Figures 4.33 and 4.34 respectively. The simulation results and harmonic spectrum for an amplitude modulation index of 1.1 are shown in Figures 4.35, 4.36 and 4.37, respectively. Simulation results are tabulated in Table 4.5.

4.5 Conclusions

For the three-phase 180° mode inverter, Tables 4.1 and 4.2 reveal that the THD of the line-to-line voltage and line-to-neutral voltage by simulation and calculation differ by a small percentage of 0.615% in both cases. The

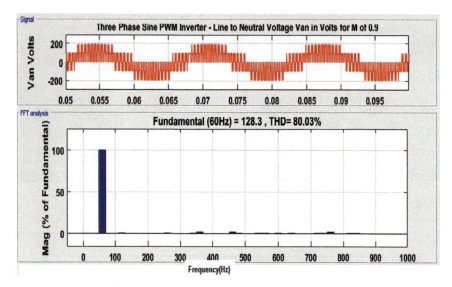

FIGURE 4.34
Three-phase sine PWM inverter: harmonic spectrum of line-to-neutral voltage for $M=0.9$.

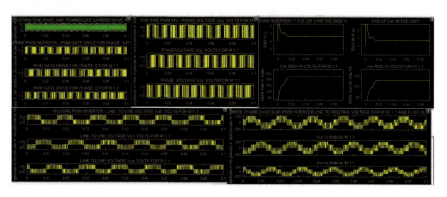

FIGURE 4.35
Three-phase sine PWM inverter simulation results for $M=1.1$.

difference in the RMS value for the line-to-line voltage and that for the line-to-neutral voltage by simulation and calculation is negligible. The simulation results in Table 4.3 for the model of three-phase 120° mode inverter reveals that the RMS values of the line-to-line and line-to-neutral voltages are close to the theoretically calculated values using the formula given in Table 4.4. The simulation result for the THD value of the line-to-line voltage differs by around 0.74% and that of the line-to-neutral voltage by around 0.776% compared with the theoretically calculated value. An interactive model for sine PWM of three-phase 180° mode inverter is also presented.

Interactive Models for DC to AC Converters

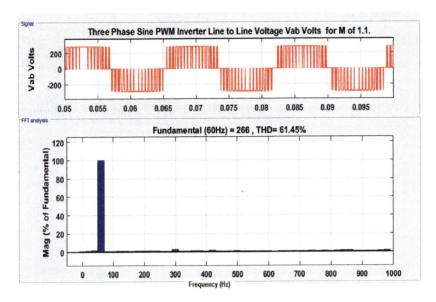

FIGURE 4.36
Three-phase sine PWM inverter: harmonic spectrum of line-to-line voltage for $M = 1.1$.

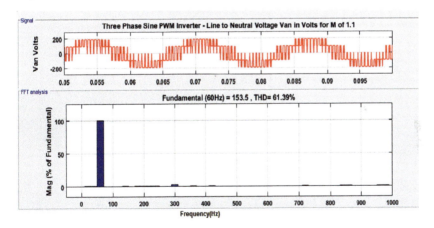

FIGURE 4.37
Three-phase sine PWM inverter: harmonic spectrum of line-to-neutral voltage for $M = 1.1$.

TABLE 4.5

Three-Phase Sine PWM Inverter: Simulation Results

Sl. No.	DC Link Voltage (V)	Amplitude Modulation Index	RMS V_{ab} (V)	THD V_{ab} (Per Unit)	RMS V_{an} (V)	THD V_{an} (Per Unit)
1	286	0.9	201.8	0.7923	116.3	0.7944
2	286	1.1	219.8	0.6134	126.6	0.6186

References

1. M.H. Rashid: *Power Electronics Circuits, Devices and Applications*, Upper Saddle River, NJ: Pearson Education, Pearson Prentice Hall, 2004.
2. I Batarseh: *Power Electronic Circuits*, Hoboken, NJ: Wiley, 2004.
3. D.W. Hart: *Introduction to Power Electronics*, Upper Saddle River, NJ: Prentice Hall, 1997.
4. N. Mohan, T.M. Undeland, and W.P. Robbins: *Power Electronics: Converters, Applications and Design*, Hoboken, NJ: Wiley, 1995.
5. B.K. Lee and M. Ehsani: "A simplified functional model for 3-phase voltage source inverter using switching function concept"; *IEEE IECON '99*; San Jose, CA, November–December 1999, Vol.1, pp. 462–467.
6. B.K. Lee and M. Ehsani: "A simplified functional simulation model for three-phase voltage source inverter using switching function concept", *IEEE Transactions on Industrial Electronics*; Vol.48, No.2, April 2001; pp. 309–321.
7. N.P.R. Iyer, V. Ramaswamy, and J. Zhu: "Simulink model for three phase sine PWM inverter fed induction motor drive", *40th International Universities Power Engineering Conference*, Cork, Ireland, September 2005, pp. 1180–1184.
8. N.P.R. Iyer: "Matlab/Simulink modules for modelling and simulation of power electronic converters and electric drives", M.E. by research thesis, University of Technology Sydney, NSW, 2006, Chapter 4.

5

Interactive Models for DC to DC Converters

5.1 Introduction

DC to DC converters are used for transforming the DC input voltage from one voltage level to the DC output voltage at another voltage level. There are three fundamental topologies for DC to DC converters, such as buck, boost and buck–boost converters. These are second-order converters, and have a semiconductor switch, a diode, one inductor, one capacitor and one load resistor and a DC input voltage source. Advanced converters, known as *fourth-order converters,* use a DC input voltage source, two inductors and two capacitors, along with one or more semiconductor switches, diodes and a load resistor. Fourth-order converters can be built by cascading any two second-order converter topologies. Additionally, Cuk, single-ended primary inductance converter (SEPIC), Zeta and Luo converters fall under the category of fourth-order converters. Buck converters find applications in switched mode power supplies, as is also the case with buck–boost converters. Boost converters are used for power factor improvement of AC–DC–AC converters used for AC drives. Cuk and SEPIC converters are used as power factor pre-regulators. These DC–DC converters are also used for charging mobile phones and laptops.

This chapter mainly deals with the interactive system modelling of fundamental topologies of second-order DC–DC converters such as buck, boost and buck–boost. The switching function concept is used to model these converters. The developed models are valid as long as the inductor current is continuous or the converter is in the continuous conduction mode (CCM) or the discontinuous conduction mode (DCM).

5.2 Buck Converter Analysis in Continuous Conduction Mode

The buck converter topology using a BJT switch is shown in Figure 5.1 [1–6]. This switch can be any one of MOSFET, IGBT or GTO. The switch is driven by an external pulse width modulation (PWM) drive whose period is T and

91

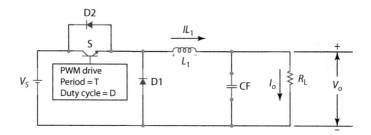

FIGURE 5.1
Buck converter.

duty cycle is D. Figure 5.2a,b shows the equivalent circuit of Figure 5.1 when the switch is ON from 0 to DT s and OFF from DT to T s, respectively. This analysis assumes CCM, that is, the inductor current is continuous or always greater than zero. Noting that the change in inductor current during the ON time and OFF time is equal or the average inductor current over one switching cycle is zero, the following equation can be derived.

$$\frac{(V_S - V_O).D.T}{L_1} + \frac{(-V_O).(1-D).T}{L_1} = 0 \tag{5.1}$$

$$\frac{V_O}{V_S} = D \tag{5.2}$$

During the switch ON time, the inductor current grows linearly and reaches a maximum value I_{L1max} and during the switch OFF period, the inductor current decays and reaches a minimum value I_{L1min}. The buck converter waveforms for the CCM operation are shown in Figure 5.3. Referring to Figure 5.3a, the average inductor current I_{L1} can be expressed as follows:

$$\begin{aligned} I_{L1} &= \frac{1}{T} * \left[\frac{T}{2} * (I_{L1max} - I_{L1min}) + T * I_{L1min} \right] \\ &= \frac{1}{2} * (I_{L1max} + I_{L1min}) \end{aligned} \tag{5.3}$$

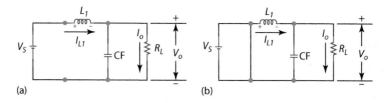

FIGURE 5.2
Buck converter equivalent circuits. (a) Switch ON for $0 < t \leq D.T$ s. (b) Switch OFF for $D.T < t \leq T$ s.

Interactive Models for DC to DC Converters

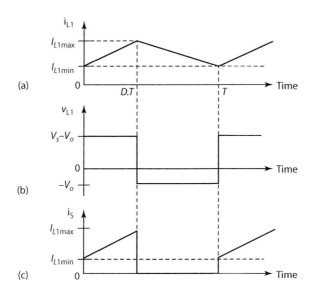

FIGURE 5.3
Buck converter waveforms in CCM. (a) Inductor current. (b) Inductor voltage. (c) Input current.

The inductor current during the switch ON and OFF intervals flows through the load resistor. The average load current I_O is given here:

$$I_O = \frac{V_O}{R_L} = \frac{1}{2} * (I_{L1max} + I_{L1min}) \tag{5.4}$$

The maximum and minimum inductor currents are given by $I_{L1max} = I_O + (\Delta i_{L1}/2)$ and $I_{L1min} = I_O - (\Delta i_{L1}/2)$, respectively. The inductor current ripple is $\Delta i_{L1} = (V_S - V_O) * D.T/L_1$. Using Equation 5.2, the maximum and minimum inductor currents are given here:

$$I_{L1max} = D.V_S * \left[\frac{1}{R_L} + \frac{(1-D)*T}{2L_1} \right]$$

$$I_{L1min} = D.V_S * \left[\frac{1}{R_L} - \frac{(1-D)*T}{2L_1} \right] \tag{5.5}$$

The critical inductance L_{1crit} required to maintain CCM is obtained by equating I_{L1min} to zero. This gives the following value for L_{1crit}.

$$L_{1crit} = \frac{(1-D)*T*R_L}{2} \tag{5.6}$$

In this derivation, fsw = $(1/T)$ is the switching frequency in hertz, L_1 is the inductor in henries and R_L is the load resistor in ohms.

5.3 Buck Converter Analysis in Discontinuous Conduction Mode

In DCM, there exists a short interval of time during which the inductor current is zero. The equivalent circuit for the DCM operation of the buck converter is shown in Figure 5.4. Figure 5.4c corresponds to the DCM operation. The inductor current and inductor voltage waveforms for DCM operation are shown in Figure 5.5. The average inductor current over one switching cycle is zero. This can be expressed as shown:

$$\frac{(V_S - V_O)*D.T}{L_1} + \frac{-V_O*(D_1-D).T}{L_1} = 0 \tag{5.7}$$

$$\frac{V_O}{V_S} = \frac{D}{D_1} \tag{5.8}$$

Referring to Figure 5.5a, using Equation 5.8, the average inductor current I_{L1} is derived as follows:

$$\begin{aligned} I_{L1} &= \frac{1}{T} * \left[\left(\frac{D_1*T}{2} \right) * \left(\frac{V_O*(D_1-D)*T}{L_1} \right) \right] \\ &= \frac{V_O*D_1*(D_1-D)*T}{2L_1} \\ &= \frac{V_O*(D_1^2 - D_1*D)*T}{2L_1} \end{aligned} \tag{5.9}$$

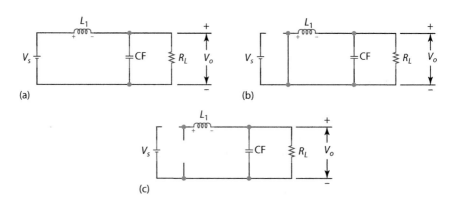

FIGURE 5.4
Buck converter equivalent circuits in DCM. (a) Switch ON. (b) Switch OFF. (c) Inductor current is zero.

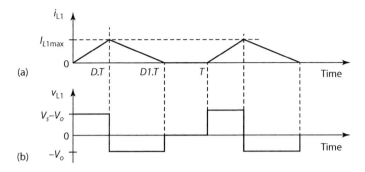

FIGURE 5.5
Buck converter waveforms in DCM. (a) Inductor current. (b) Inductor voltage.

Because the average inductor current I_{L1} and the load current I_O are equal, we have the following:

$$\frac{V_O}{R_L} = \frac{V_O * (D_1^2 - D_1 * D) * T}{2L_1} \tag{5.10}$$

Solving Equation 5.10, the value of D_1 and the voltage gain V_O/V_S for DCM are given here:

$$D_1^2 - D_1.D - \frac{2.L_1}{R_L.T} = 0$$

$$D_1 = \frac{D}{2} \pm \sqrt{\frac{D^2}{4} + \frac{2.L_1}{R_L.T}} \tag{5.11}$$

$$M = \frac{V_O}{V_S} = \frac{D}{\frac{D}{2} \pm \sqrt{\frac{D^2}{4} + \frac{2.L_1}{R_L.T}}} \tag{5.12}$$

Equations 5.11 and 5.12 give two values for D_1 and M, of which only one is valid.

5.4 Model of Buck Converter in CCM and DCM

Models for the buck converter are available in references [7–13]. The switching function concept is used to model the buck converter [7–13]. Buck converter model parameters are shown in Table 5.1. The development of the model is given as follows:

TABLE 5.1

Buck Converter Model Parameters

Sl. No.	Input Voltage V_S (V)	Switching Period T (s)	Duty Cycle D	Inductor L_1 (H)	Filter Capacitor C_F (F)	Load Resistor R_L (Ω)	Remarks
1	12	0.2E–3	5/12	200E–6	47E–6	2.5	CCM
2	12	0.2E–3	5/12	50E–6	470E–6	2.5	DCM

Referring to Figures 5.2 and 5.4, the following equations can be derived for CCM and DCM operation.

$$
\begin{aligned}
i_{L1\_dot} &= \frac{(V_S - V_O)}{L_1} \quad \text{for} \quad 0 \le t \le D.T \\
&= \frac{(-V_O)}{L_1} \quad \text{for} \quad D.T \le t \le T
\end{aligned}
\tag{5.13}
$$

Now, for the switch S, the switching function SF is defined as follows for CCM operation:

$$
\begin{aligned}
SF &= 1 \quad \text{for} \quad 0 \le t \le D.T \\
&= 0 \quad \text{for} \quad D.T \le t \le T
\end{aligned}
\tag{5.14}
$$

For the inductor current i_{L1}, another switch function SF_IL1 is defined:

$$
\begin{aligned}
SF\_IL1 &= 1 \quad \text{for} \quad i_{L1} \succ 0 \\
&= 0 \quad \text{for} \quad i_{L1} \succ 0
\end{aligned}
\tag{5.15}
$$

Using Equations 5.14 and 5.15 in Equation 5.13, we have the following:

$$
i_{L1\_dot} = \frac{(SF * V_S - SF\_IL1 * V_O)}{L_1}
\tag{5.16}
$$

The inductor current flows through the load resistor during switch ON and OFF in CCM and in DCM until it falls to zero. The output voltage equation is formulated by observation of equivalent circuits for CCM and DCM (Figures 5.2 and 5.4), as follows:

$$
I_{CF} = I_{L1} - I_O
\tag{5.17}
$$

$$
v_{O\_dot} = \left(\frac{SF\_IL1 * I_{L1}}{C_F} - \frac{V_O}{R_L * C_F} \right)
\tag{5.18}
$$

Interactive Models for DC to DC Converters

$$I_O = \frac{V_O}{R_L} \tag{5.19}$$

In Equations 5.16 and 5.18, i_{L1_dot} and v_{O_dot} represent di_{L1}/dt and dv_O/dt, respectively. In Equation 5.17, I_{CF} is the average filter current and I_O is the load current.

The interactive model of the buck converter suitable for CCM and DCM operation is shown in Figure 5.6 with a dialogue box. The model subsystem is shown in Figure 5.7. The data in the dialogue box in Figure 5.6 corresponds to the CCM operation given in Table 5.1. The appropriate data for DCM operation given in Table 5.1 are entered in the dialogue box if DCM operation is required.

The switch function SF for the switch defined in Equation 5.14 is generated using the Pulse Generator block, whose amplitude is 1, pulse width 'd*100' (% of period) and switching period '1/(fsw)', where 'd' is the duty ratio and 'fsw' is the frequency of switching. The switch function for the inductor current SF_IL1 defined in Equation 5.15 is generated using the embedded MATLAB® function, whose input is IL1 and output is SF_IL1 as shown in Figure 5.7. Equations 5.16, 5.18 and 5.19 are solved using another embedded MATLAB function and integrators, as shown in Figure 5.7. The source code for the two embedded MATLAB functions are given here:

```
function [IL1_dot, VO_dot, IO] = fcn(SF,IL1,VO,VS,L1,CF,RL,SF_IL1)
IL1_dot = -SF_IL1*VO/(L1) + SF*VS/(L1);
VO_dot = SF_IL1*IL1/(CF) - VO/(RL*CF);
IO = VO/(RL);

function SF_IL1 = fcn(IL1)
if IL1 <= 0
    SF_IL1 = 0;
else
    SF_IL1 = 1;
End
```

The embedded MATLAB outputs v_{o_dot}, i_{L1_dot} are integrated using Integrator blocks to obtain the required outputs and are used as feedback inputs along with SF_IL1 to solve Equations 5.16, 5.18 and 5.19.

5.4.1 Simulation Results

Simulation of the buck converter for CCM and DCM operations was carried out using Simulink® [14]. The ode23t (mod. stiff/trapezoidal) solver was used. The data shown in Table 5.1 were used for simulation. The simulation results for the buck converter in CCM are shown in Figure 5.8 and those for DCM are shown in Figure 5.9. Simulation results and calculated values are tabulated in Tables 5.2 and 5.3.

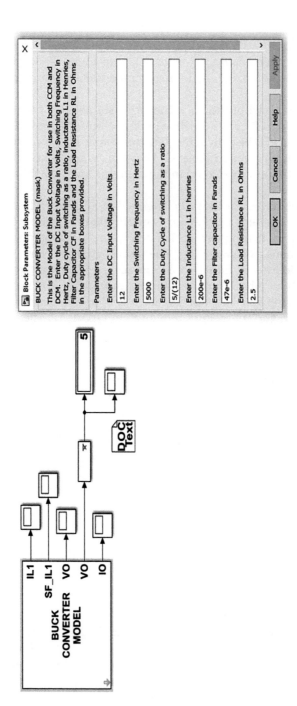

FIGURE 5.6
Model of buck converter in CCM and DCM.

Interactive Models for DC to DC Converters 99

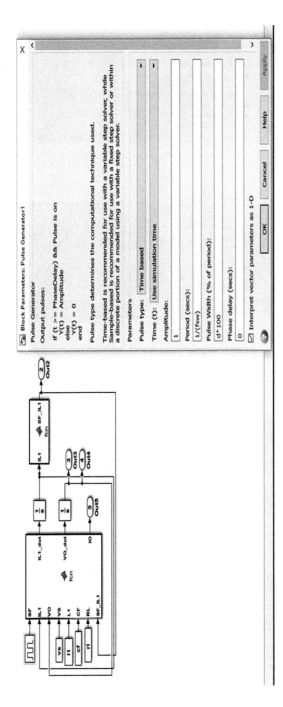

FIGURE 5.7
Buck converter model subsystem.

100   Power Electronic Converters

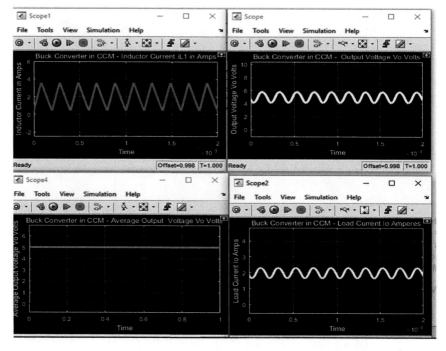

FIGURE 5.8
Buck converter in CCM: simulation results.

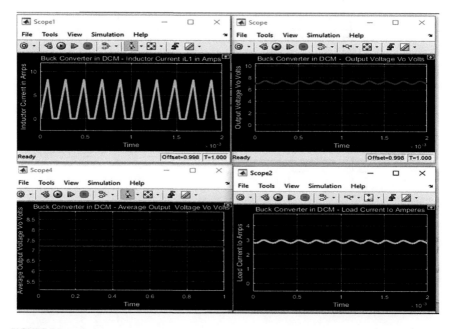

FIGURE 5.9
Buck converter in DCM: simulation results.

Interactive Models for DC to DC Converters

TABLE 5.2

Buck Converter: Simulation Results

Sl. No.	Output Voltage V_O (V)	Minimum Inductor Current (A)	Maximum Inductor Current (A)	Load Current (A)	Remarks
1	5	0.425	3.594	2	CCM
2	7.2	0	8.1	2.88	DCM

TABLE 5.3

Buck Converter: Calculated Values

Sl. No.	Output Voltage V_O (V)	Minimum Inductor Current (A)	Maximum Inductor Current (A)	Load Current (A)	Remarks
1	5	0.54	3.46	2	CCM
2	7.44	0	7.6	2.976	DCM

5.5 Boost Converter Analysis in Continuous Conduction Mode

The boost converter topology using a semiconductor switch is shown in Figure 5.10 [1–6]. The switch is driven by an external PWM drive whose period is T and duty cycle is D. Figure 5.11a,b shows the equivalent circuit of Figure 5.10 in the CCM when the switch is ON from 0 to DT s and OFF from DT to T s respectively. The inductor current is always greater than zero in the CCM analysis. Noting that the change in inductor current during the ON period and OFF period are equal or the average inductor current over one switching cycle is zero, the following equations can be derived:

$$\left(\frac{V_S}{L_1}\right)*D.T + \left[\frac{(V_S - V_O)}{L_1}\right]*(1-D).T = 0 \tag{5.20}$$

$$\frac{V_O}{V_S} = \frac{1}{(1-D)} \tag{5.21}$$

During the switch ON period, the inductor current grows linearly and reaches a maximum value I_{L1max}, and during the switch OFF period, the inductor current decays and reaches a minimum value I_{L1min}. The relevant waveforms for CCM operation of the boost converter are shown in Figure 5.12. Neglecting switching losses, input and output power are equal. The inductor current is derived as follows:

$$V_S * I_{L1} = \frac{V_O^2}{R_L} \tag{5.22}$$

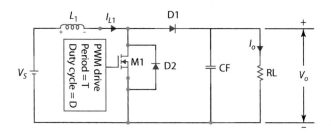

FIGURE 5.10
Boost converter.

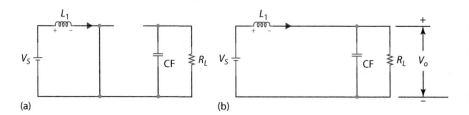

FIGURE 5.11
Boost converter equivalent circuits. (a) Switch ON for $0 < t \leq D.T$. (b) Switch OFF for $D.T < t \leq T$.

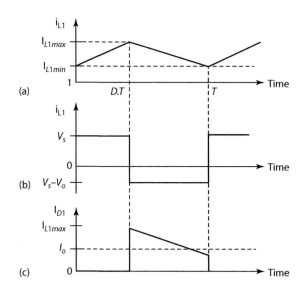

FIGURE 5.12
Boost converter waveforms in CCM: (a) inductor current; (b) inductor voltage; and (c) input current.

Interactive Models for DC to DC Converters 103

Using Equations 5.21 and 5.22, the average inductor current I_{L1} is given as follows:

$$I_{L1} = \frac{V_S}{(1-D)^2 * R_L} \tag{5.23}$$

The maximum and minimum inductor current are derived as follows:

$$I_{L1max} = \left[\frac{V_S}{(1-D)^2 * R_L} + \frac{V_S * D.T}{2.L_1}\right] \tag{5.24}$$

$$I_{L1min} = \left[\frac{V_S}{(1-D)^2 * R_L} - \frac{V_S * D.T}{2.L_1}\right] \tag{5.25}$$

The critical inductance L_{1crit} required to maintain CCM is obtained by equating I_{L1min} to zero. This gives the following value for L_{1crit}:

$$L_{1crit} = \frac{R_L * T}{2} * (1-D)^2 * D \tag{5.26}$$

If the inductor value of L_1 is below L_{1crit}, then the converter operates in DCM.

5.6 Boost Converter Analysis in Discontinuous Conduction Mode

In DCM, there exists a short interval of time during which the inductor current is zero. The equivalent circuit for DCM operation of the buck converter is shown in Figure 5.13. Figure 5.13c corresponds to DCM operation. The relevant voltage and current waveforms for DCM operation are shown in Figure 5.14. The average inductor current over one switching cycle is zero. This can be expressed as shown here:

$$\left(\frac{V_S}{L_1}\right) * D.T + \left[\frac{(V_S - V_O)}{L_1}\right] * (D_1 - D).T = 0 \tag{5.27}$$

$$\frac{V_O}{V_S} = M = \frac{D_1}{(D_1 - D)} \tag{5.28}$$

Referring to Figure 5.14c, the average diode current is equal to the load current. Thus, I_O is expressed as follows:

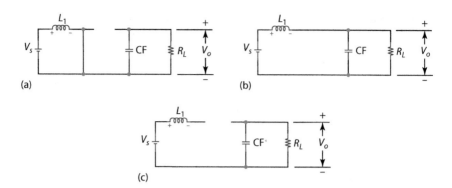

FIGURE 5.13
Boost converter equivalent circuits in DCM. (a) Switch ON. (b) Switch OFF. (c) Inductor and diode current are zero.

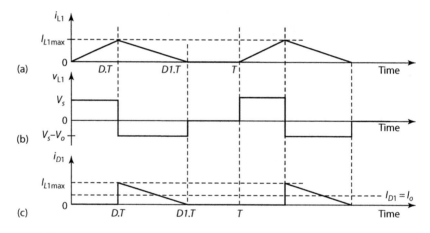

FIGURE 5.14
Boost converter waveforms in DCM. (a) Inductor current. (b) Inductor voltage. (c) Diode current.

$$I_O = \frac{V_O}{R_L} = \frac{(D_1 - D) * I_{L1max}}{2} = \frac{V_S * (D_1 - D) * D * T}{2.L_1} \quad (5.29)$$

$$\frac{V_O}{V_S} = M = \frac{(D_1 - D) * D * T * R_L}{2.L_1} \quad (5.30)$$

Using Equations 5.28 and 5.30 and solving for M, we have the following:

$$\frac{V_O}{V_S} = M = \frac{1}{2} \pm \sqrt{\frac{1}{4} + \frac{D^2 * T * R_L}{2.L_1}} \quad (5.31)$$

Interactive Models for DC to DC Converters

$$D_1 = \frac{M * D}{(M - 1)} \qquad (5.32)$$

Equation 5.31 gives two values for M, of which one is valid.

5.7 Model of Boost Converter in CCM and DCM

Models of the boost converter are available in references [7–13]. The switching function concept is used to model the boost converter [7–13]. Boost converter model parameters are shown in Table 5.4. The development of the model is given here:

Referring to Figures 5.11 and 5.13, the following equations can be derived for CCM and DCM operation of boost converters.

$$
\begin{aligned}
i_{L1\_dot} &= \frac{(V_S)}{L_1} \quad \text{for} \quad 0 \le t \le D.T \\
&= \frac{(V_S - V_O)}{L_1} \quad \text{for} \quad D.T \le t \le T
\end{aligned}
\qquad (5.33)
$$

Now, for the switch S, switching function SF has already been defined in Equation 5.14 for CCM operation. The inverse switch function SF_BAR is defined as follows:

$$
\begin{aligned}
SF\_BAR &= 0 \quad \text{for} \quad 0 \le t \le D.T \\
&= 1 \quad \text{for} \quad D.T \le t \le T
\end{aligned}
\qquad (5.34)
$$

$$SF\_BAR = (1 - SF) \qquad (5.35)$$

For the inductor current i_{L1}, another switch function SF_IL1 has already been defined in Equation 5.15.

Using Equations 5.14, 5.15 and 5.35 in Equation 5.33, we have the following:

TABLE 5.4

Boost Converter Model Parameters

Sl. No.	Input Voltage V_S (V)	Switching Frequency fsw (Hz)	Duty Cycle D	Inductor L_1 (H)	Filter Capacitor C_F (F)	Load Resistor R_L (Ω)	Remarks
1	12	25E3	0.6	960E–6	90E–6	50	CCM
2	12	25E3	0.6	20E–6	90E–6	50	DCM

$$i_{L1\_dot} = \frac{\left[V_S - \text{SF\_IL1}*(1-\text{SF})*V_O\right]}{L_1} \tag{5.36}$$

The inductor current flows through the load resistor during the switch OFF period in CCM and in DCM until it falls to zero. The output voltage equation is formulated by observation of equivalent circuits for CCM and DCM (Figures 5.11 and 5.13), as follows:

$$I_{CF} = (1-\text{SF})*I_{L1} - I_O \tag{5.37}$$

$$v_{O\_dot} = \left(\frac{\text{SF\_IL1}*(1-\text{SF})*I_{L1}}{C_F} - \frac{V_O}{R_L*C_F}\right) \tag{5.38}$$

In Equations 5.36, 5.37 and 5.38, i_{L1_dot}, I_{CF}, I_O and v_{O_dot} are as defined in Section 5.4. I_O is defined in Equation 5.19.

The interactive model of the boost converter suitable for CCM and DCM operation is given in Figure 5.15 with a dialogue box. The model subsystem with a dialogue box is shown in Figure 5.16. The data shown in the dialogue box is for CCM operation of the boost converter, given in Table 5.4. For DCM operation, the relevant data from Table 5.4 have to be re-entered in the dialogue box.

The switch function for the switch SF defined in Equation 5.14 and that for inductor current SF_IL1 defined in Equation 5.15 are generated as explained in Section 5.4. Equations 5.36, 5.38 and 5.19 are solved using the embedded MATLAB function and the source code is written as explained in Section 5.4. The embedded MATLAB outputs v_{o_dot}, i_{L1_dot} are integrated using Integrator blocks to obtain the required outputs and are used as feedback inputs along with SF_IL1 to solve Equations 5.36, 5.38 and 5.19.

5.7.1 Simulation Results

Simulation of the boost converter for CCM and DCM operations were carried out using Simulink [14]. The ode23t (mod. stiff/trapezoidal) solver was used. The data shown in Table 5.4 were used for the simulation. The simulation results for the boost converter in CCM are shown in Figure 5.17 and those for DCM are shown in Figure 5.18. Simulation results and calculated values are tabulated in Tables 5.5 and 5.6.

5.8 Buck–Boost Converter Analysis in Continuous Conduction Mode

The buck–boost converter topology using a semiconductor switch is shown in Figure 5.19 [1–6]. The switch is driven by an external PWM drive whose period is T and duty cycle is D. Figure 5.20a,b shows the equivalent

Interactive Models for DC to DC Converters 107

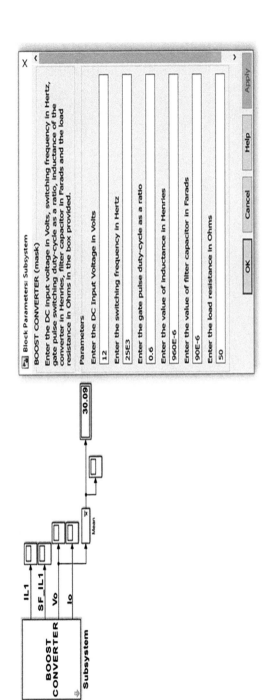

FIGURE 5.15
Boost converter in CCM and DCM.

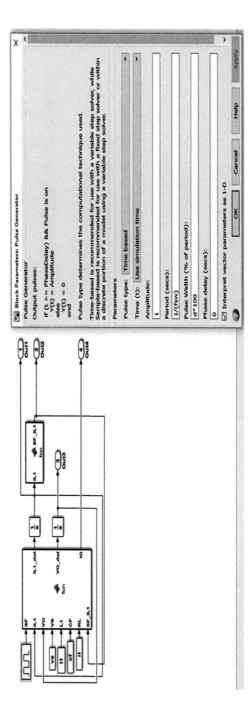

FIGURE 5.16
Boost converter model subsystem.

Interactive Models for DC to DC Converters

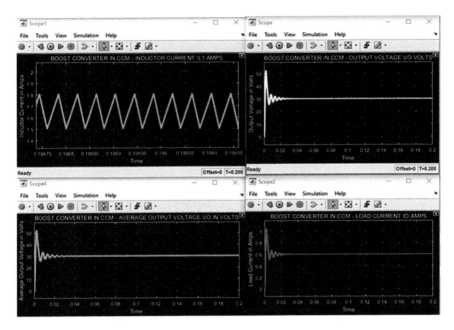

FIGURE 5.17
Boost converter in CCM: simulation results.

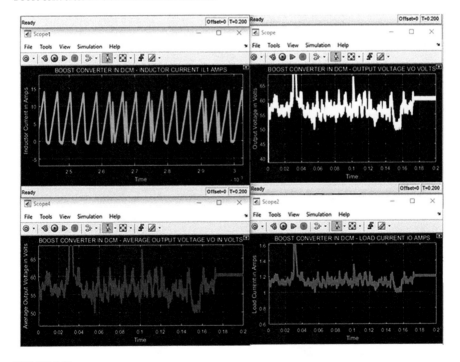

FIGURE 5.18
Boost converter in DCM: simulation results.

TABLE 5.5
Boost Converter: Simulation Results

Sl. No.	Output Voltage V_O (V)	Minimum Inductor Current (A)	Maximum Inductor Current (A)	Load Current (A)	Remarks
1	30.09	1.5	1.8	0.6	CCM
2	60	0	14	1.2	DCM

TABLE 5.6
Boost Converter: Calculated Values

Sl. No.	Output Voltage V_O (V)	Minimum Inductor Current (A)	Maximum Inductor Current (A)	Load Current (A)	Remarks
1	30	1.35	1.65	0.6	CCM
2	57.26	0	14.4	1.145	DCM

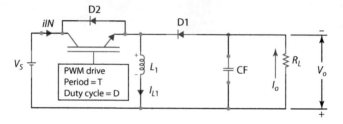

FIGURE 5.19
Buck–boost converter.

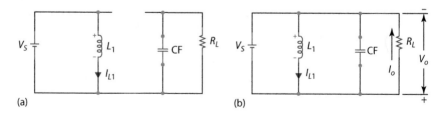

FIGURE 5.20
Buck–boost converter equivalent circuits in CCM. (a) Switch ON for $0 < t \leq D.T$. (b) Switch OFF for $D.T < t \leq T$.

circuit of Figure 5.19 in the CCM when the switch is ON from 0 to DT s and OFF from DT to T s respectively. The inductor current is always greater than zero in the CCM analysis. Noting that the change in inductor current during the ON period and OFF period are equal or the average inductor current over one switching cycle is zero, the following equations can be derived:

Interactive Models for DC to DC Converters

$$\left(\frac{V_S}{L_1}\right)*D*T + \left(\frac{V_O}{L_1}\right)*(1-D)*T = 0 \quad (5.39)$$

$$\frac{V_O}{V_S} = \frac{-D}{(1-D)} \quad (5.40)$$

The minus sign in Equation 5.40 indicates that the polarity of V_O is negative at the top and positive at the bottom terminal. During the switch ON period, the inductor current grows linearly and reaches a maximum value I_{L1max}, and during the switch OFF period, the inductor current decays and reaches a minimum value I_{L1min}. The relevant waveforms for CCM operation of the buck–boost converter are shown in Figure 5.21. Neglecting switching losses, input and output power are equal. The inductor current is derived as follows:

$$V_S * I_{IN} = \frac{V_O^2}{R_L} \quad (5.41)$$

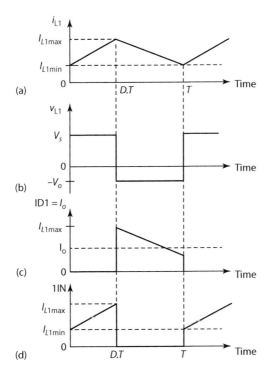

FIGURE 5.21
Buck–boost converter waveforms in CCM. (a) Inductor current. (b) Inductor voltage. (c) Diode current. (d) Input current.

$$I_{IN} = D * I_{L1}$$ (5.42)

$$I_{L1} = \frac{V_O^2}{V_S * D * R_L}$$ (5.43)

Using Equations 5.40 and 5.43, the average inductor current I_{L1} and the maximum and minimum values of the inductor current can be expressed as follows:

$$I_{L1} = \frac{V_S * D}{(1-D)^2 * R_L}$$ (5.44)

$$I_{L1\,max} = \left[\frac{V_S * D}{(1-D)^2 * R_L} + \frac{V_S * D * T}{2.L_1} \right]$$ (5.45)

$$I_{L1\,min} = \left[\frac{V_S * D}{(1-D)^2 * R_L} - \frac{V_S * D * T}{2.L_1} \right]$$ (5.46)

The critical inductance, or minimum value of inductance of the buck–boost converter to maintain CCM operation, is obtained by setting Equation 5.46 to zero, which results in the following expression:

$$L_{1crit} = \frac{(1-D)^2 * T * R_L}{2}$$ (5.47)

5.9 Buck–Boost Converter Analysis in the Discontinuous Conduction Mode

In the DCM, there exists a short interval of time during which the inductor current is zero. The equivalent circuit for DCM operation of the buck–boost converter is shown in Figure 5.22. Figure 5.22c corresponds to DCM operation. The relevant voltage and current waveforms for DCM operation are shown in Figure 5.23. The average inductor current over one switching cycle is zero. This can be expressed as follows:

$$\left(\frac{V_S}{L_1} \right) * D.T + \left(\frac{-V_O}{L_1} \right) * (D_1 - D).T = 0$$ (5.48)

Interactive Models for DC to DC Converters

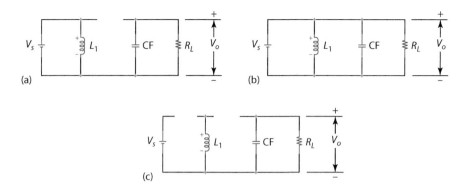

FIGURE 5.22
Buck–boost converter equivalent circuits in DCM. (a) Switch ON; (b) Switch OFF. (c) Inductor and diode current are zero.

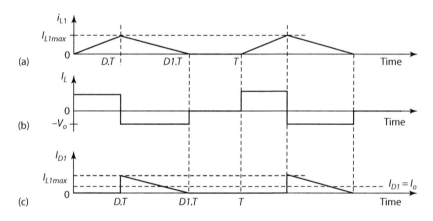

FIGURE 5.23
Buck–boost converter waveforms in DCM. (a) Inductor current. (b) Inductor voltage. (c) Diode current.

$$\frac{V_O}{V_S} = M = \frac{D}{(D_1 - D)} \tag{5.49}$$

Referring to Figure 5.23c, the average diode current is equal to the load current. Thus, I_O is expressed as follows.

$$I_O = \frac{V_O}{R_L} = \frac{(D_1 - D) * I_{L1\max}}{2} = \frac{V_S * (D_1 - D) * D * T}{2.L_1} \tag{5.50}$$

$$\frac{V_O}{V_S} = M = \frac{(D_1 - D) * D * T * R_L}{2.L_1} \tag{5.51}$$

Using Equations 5.49 and 5.51 and solving for M, we have the following:

$$\frac{V_O}{V_S} = M = D * \sqrt{\frac{T*R_L}{2.L_1}}$$

(5.52)

$$D_1 = \left(D + \frac{D}{M}\right)$$

(5.53)

5.10 Models of Buck–Boost Converter in CCM and DCM

Models of the buck–boost converter are available in the literature [13]. The switching function concept is used to model the buck–boost converter [7–13]. The buck–boost converter model parameters are shown in Table 5.7. The development of the model is given here.

Referring to Figures 5.20 and 5.22, the following equations can be derived for CCM and DCM operation of the buck–boost converter:

$$i_{L1\_dot} = \frac{(V_S)}{L_1} \quad \text{for} \quad 0 \le t \le D.T$$

$$= \frac{(-V_O)}{L_1} \quad \text{for} \quad D.T \le t \le T$$

(5.54)

Now, for the switch S, the switching function SF has already been defined in Equation 5.14 for CCM operation. The inverse switch function SF_BAR has been defined in Equations 5.34 and 5.35.

For the inductor current i_{L1}, another switch function SF_IL1 has already been defined in Equation 5.15.

Using Equations 5.14, 5.15 and 5.35 in Equation 5.54, we have the following:

$$i_{L1\_dot} = \frac{\left[SF * V_S - SF\_IL1 * (1-SF) * V_O\right]}{L_1}$$

(5.55)

TABLE 5.7

Buck–Boost Converter Model Parameters

Sl. No.	Input Voltage V_S(V)	Switching Period T (s)	Duty Cycle D	Inductor L_1 (H)	Filter Capacitor C_F (F)	Load Resistor R_L (Ω)	Remarks
1	24	5.00E−05	0.4	1.00E−04	4.00E−04	5	CCM
2	24	5.00E−05	0.6	1.00E−04	4.00E−04	5	CCM
3	24	5.00E−05	0.4	2.30E−05	4.00E−04	5	DCM

Interactive Models for DC to DC Converters

The inductor current flows through the load resistor during the switch OFF period in CCM and in DCM until it falls to zero. The output voltage equation is formulated by observation of equivalent circuits for CCM and DCM (Figures 5.20 and 5.22), as follows:

$$I_{CF} = (1-SF) * I_{L1} - I_O \qquad (5.56)$$

$$v_{O\_dot} = \left(\frac{SF\_IL1 * (1-SF) * I_{L1}}{C_F} - \frac{V_O}{R_L * C_F} \right) \qquad (5.57)$$

In Equations 5.55, 5.56 and 5.57, i_{L1_dot}, I_{CF}, I_O and v_{O_dot} are as defined in Section 5.4. I_O is defined in Equation 5.19.

The interactive model of the buck–boost converter suitable for CCM and DCM operation is given in Figure 5.24 with a dialogue box. The model subsystem with a dialogue box is shown in Figure 5.25.

The data shown in the dialogue box is for CCM operation of the boost converter given in Table 5.7. For DCM operation, the relevant data from Table 5.7 have to be re-entered in the dialogue box.

The switch function for the switch SF defined in Equation 5.14 and that for the inductor current SF_IL1 defined in Equation 5.15 are generated as explained in Section 5.4. Equations 5.55, 5.57 and 5.19 are solved using the embedded MATLAB function and the source code is written as explained in Section 5.4. The embedded MATLAB outputs v_{o_dot}, i_{L1_dot} are integrated using Integrator blocks to obtain the required outputs and are used as feedback inputs along with SF_IL1 to solve Equations 5.55, 5.57 and 5.19.

5.10.1 Simulation Results

Simulation of the buck–boost converter for CCM and DCM operations was carried out via Simulink [14]. The ode23t (mod. stiff/trapezoidal) solver was used. The data shown in Table 5.7 are used for the simulation. The simulation results for the buck–boost converter in CCM are shown in Figures 5.26 and 5.27 and those for DCM are shown in Figure 5.28. Simulation results and calculated values are tabulated in Tables 5.8 and 5.9, respectively.

5.11 Conclusions

Interactive models are developed for the second-order DC to DC converters such as the buck, boost and the buck–boost topology. One model is suitable to analyse both CCM and DCM operation. The models developed are system models. The embedded MATLAB function is used to develop system models

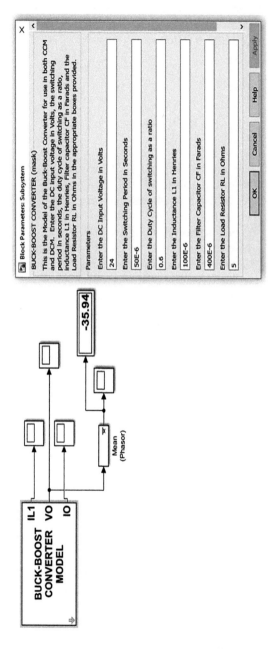

FIGURE 5.24
Buck–boost converter model in CCM and DCM.

Interactive Models for DC to DC Converters 117

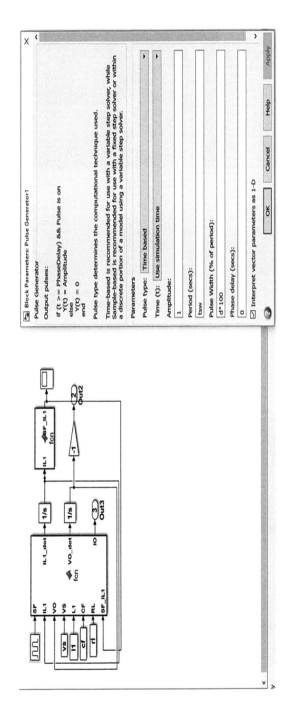

FIGURE 5.25
Buck–boost converter model subsystem.

118   *Power Electronic Converters*

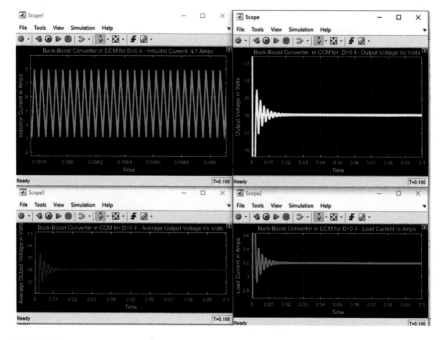

FIGURE 5.26
Buck–boost converter in CCM: simulation results for $D = 0.4$.

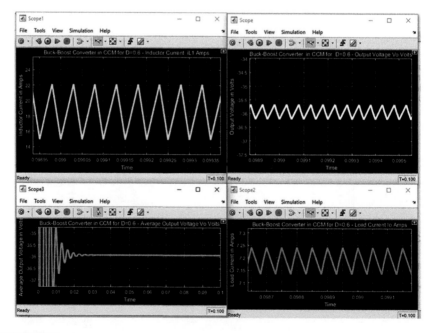

FIGURE 5.27
Buck–boost converter in CCM: simulation results for $D = 0.6$.

Interactive Models for DC to DC Converters

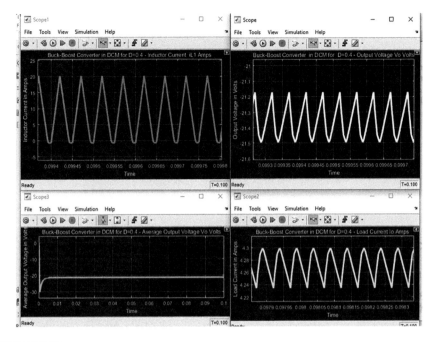

FIGURE 5.28
Buck–boost converter in DCM: simulation results for $D = 0.4$.

TABLE 5.8

Buck–Boost Converter: Simulation Results

Sl. No.	Output Voltage V_O (V)	Minimum Inductor Current (A)	Maximum Inductor Current (A)	Load Current (A)	Duty Cycle	Remarks
1	−16	3.05	7.99	3.2	0.4	CCM
2	−35.94	15	22	7.19	0.6	CCM
3	−21.35	0	20	4.27	0.4	DCM

TABLE 5.9

Buck–Boost Converter: Calculated Values

Sl. No.	Output Voltage V_O (V)	Minimum Inductor Current (A)	Maximum Inductor Current (A)	Load Current (A)	Duty Cycle	Remarks
1	−16	2.933	7.733	3.2	0.4	CCM
2	−36	14.4	21.6	7.2	0.6	CCM
3	−22.38	0	20.869	4.476	0.4	DCM

for all these DC to DC converters. The simulation results indicate that the deviation from theoretically calculated values using formulae are negligible for all three DC to DC converters.

References

1. M.H. Rashid: *Power Electronics Circuits, Devices and Applications*, Upper Saddle River, NJ: Pearson Education, Pearson Prentice Hall, 2004.
2. I. Batarseh: *Power Electronic Circuits*, Hoboken, NJ: Wiley, 2004.
3. D.W. Hart: *Introduction to Power Electronics*, Upper Saddle River, NJ: Prentice Hall, 1997.
4. N. Mohan, T.M. Undeland, and W.P. Robbins: *Power Electronics: Converters, Applications and Design*, Hoboken, NJ: Wiley, 1995; Ch.4; pp. 61–76.
5. B. Choi: *Pulse Width Modulated DC to DC Power Conversion: Circuit, Dynamics and Control Designs*, Piscataway, NJ: IEEE Press – Wiley, 2013; pp. 93–143.
6. F.L. Luo and H. Ye: *Power Electronics*, Boca Raton, FL: CRC Press, 2013.
7. V.F. Pires and J.F.A. Silva: "Teaching nonlinear modelling, simulation and control of electronic power converters using MATLAB/SIMULINK", *IEEE Transactions on Education*; Vol.45, No.3, August 2002; pp. 253–256.
8. B. Baha: "Modelling of resonant switched-mode converters using SIMULINK", *IEE Proceedings, Electric Power Applications*; Vol.145, No.3, May 1998; pp. 159–163.
9. B. Baha: "Simulation of switched-mode power electronic circuits", *IEE International Conference on Simulation*, United Kingdom, 1998; pp. 209–214.
10. A.N. Melendez, J.D. Gandoy, C.M. Penalver, and A. Lago: "A new complete non-linear simulation model of a buck DC-DC converter", *IEEE-ISIE'99*, Slovenia, 1999; pp. 257–261.
11. H.Y. Kanaan and K. Al-Haddad: "Modeling and simulation of DC-DC power converters in CCM and DCM using the switching functions approach: Applications to the buck and Cuk converters", *IEEE-PEDS 2005*, Malaysia, November–December, 2005; pp. 468–473.
12 H.Y. Kanaan, K. Al-Haddad, and F. Fnaiech: "Switching-function-based modeling and control of a SEPIC power factor correction circuit operating in continuous and discontinuous current modes", *IEEE International Conference on Industrial Technology*, Tunisia, 2004; pp. 431–437.
13. N.P.R. Iyer: "MATLAB/SIMULINK modules for modelling and simulation of power electronic converters and electric drives", M.E. by research thesis, University of Technology Sydney, NSW, Australia, 2006, Chapter 5.
14. The Mathworks Inc.: "MATLAB/Simulink Release Notes", R2016b, 2016.

6

Interactive Models for AC to AC Converters

6.1 Introduction

There are two categories of AC to AC converters. One is the cycloconverter, which converts the AC input voltage with a given frequency and peak value to a variable amplitude AC output voltage at a desired frequency, different from the frequency of the AC input voltage. The modelling of cycloconverters is not discussed in this chapter. The AC to AC converters used to vary the root mean square (RMS) value of the load voltage at a constant frequency that is the same as the frequency of the AC input voltage are known as *AC voltage controllers* or *AC regulators*. The voltage control is accomplished by phase control or on–off control. A single-phase AC controller uses a pair of silicon-controlled rectifiers (SCRs) connected back-to-back or antiparallel between the supply and the load. Several configurations for three-phase AC controllers are available employing three single-phase AC controllers. This chapter describes the system modelling of two predominant three-phase AC controller configurations, namely, three-phase AC controllers with back-to-back SCRs in series with the AC lines with star-connected resistive load and isolated neutral; and the other, the three-phase AC controllers with back-to-back SCRs in series with resistive load connected in delta to the AC lines.

6.2 Analysis of a Fully Controlled Three-Phase Three-Wire AC Voltage Controller with Star-Connected Resistive Load and Isolated Neutral

The three-phase three-wire AC voltage controller with back-to-back antiparallel SCRs and star-connected resistive load with isolated neutral is shown in Figure 6.1 [1–4].

By varying the firing angle α, the RMS voltage across the load can be varied. The firing sequence is T1, T2, T3, T4, T5 and T6. This is derived as follows:

The three-phase line-to-neutral input voltages v_{AN}, v_{BN} and v_{CN} are defined in Equation 6.1:

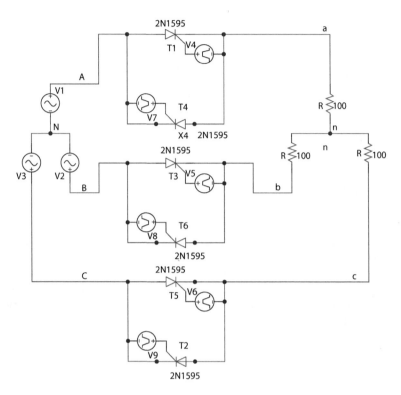

FIGURE 6.1
Three-phase AC back-to-back connected thyristor controller.

$$v_{AN} = \sqrt{2}*V_S*\sin(\omega.t)$$
$$v_{BN} = \sqrt{2}*V_S*\sin\left(\omega.t - \frac{2\pi}{3}\right) \qquad (6.1)$$
$$v_{CN} = \sqrt{2}*V_S*\sin\left(\omega.t + \frac{2\pi}{3}\right)$$

The three-phase instantaneous line-to-line input voltages are given here:

$$v_{AB} = \sqrt{6}*V_S*\sin\left(\omega.t + \frac{\pi}{6}\right)$$
$$v_{BC} = \sqrt{6}*V_S*\sin\left(\omega.t - \frac{\pi}{2}\right) \qquad (6.2)$$
$$v_{CA} = \sqrt{6}*V_S*\sin\left(\omega.t - \frac{7\pi}{6}\right)$$

The RMS line-to-neutral voltage across the star-connected resistive load for various ranges of firing angle α are given here [1]:

For $0 \leq \alpha < 60°$

$$V_O = V_{an\_rms} = \sqrt{6} * V_S * \sqrt{\left[\frac{1}{6} - \frac{\alpha}{4\pi} + \frac{\sin(2\alpha)}{8\pi}\right]} \quad (6.3)$$

For $60° \leq \alpha < 90°$

$$V_O = V_{an\_rms} = \sqrt{6} * V_S * \sqrt{\left[\frac{1}{12} - \frac{3*\sin(2\alpha)}{16\pi} + \frac{\sqrt{3}*\cos(2\alpha)}{16\pi}\right]} \quad (6.4)$$

For $90° \leq \alpha < 150°$

$$V_O = V_{an\_rms} = \sqrt{6} * V_S * \sqrt{\left[\frac{5}{24} - \frac{\alpha}{4\pi} - \frac{\sin(2\alpha)}{16\pi} + \frac{\sqrt{3}*\cos(2\alpha)}{16\pi}\right]} \quad (6.5)$$

For $0 \leq \alpha < 60°$, immediately before firing of T1, two thyristors conduct. When T1 is fired, three thyristors conduct. When the thyristor current reverses, the particular thyristor turns OFF. For $60° \leq \alpha < 90°$, only two thyristors conduct at any time. For $90° \leq \alpha < 150°$, although two thyristors conduct at any time, there are periods when *no* thyristors are ON. The output voltage becomes zero for α of 150°.

A PSIM [5] circuit model of this three-phase AC controller with star-connected resistive load and isolated neutral is shown in Figure 6.2 and the simulation result for a firing angle α of π/6 rad is shown in Figure 6.3.

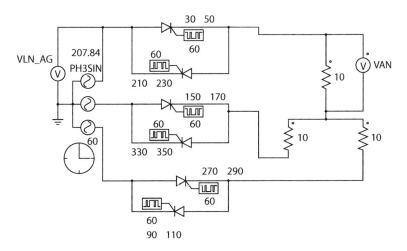

FIGURE 6.2
Three-phase AC controller with star-connected resistive load.

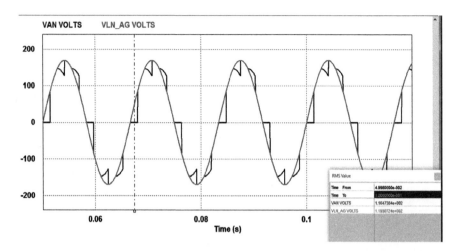

FIGURE 6.3
Simulation results for SCR three-phase AC controller with star-connected resistive load for a firing angle α of 30°.

The line-to-neutral RMS input voltage is 120 V, with a frequency of 60 Hz. The line-to-neutral RMS output voltage V_{an} is 116.47 V by simulation, which closely agrees with the result obtained by using Equation 6.3.

6.2.1 Modelling of a Fully Controlled Three-Phase Three-Wire AC Voltage Controller with Star-Connected Resistive Load and Isolated Neutral

This section describes the modelling of the three-phase AC controller shown in Figure 6.1 [6]. The model for a firing angle α of $\pi/6$ rad is shown in Figure 6.4. The various dialogue boxes are shown in Figure 6.5 and the various subsystems in Figure 6.6a–d. The various subsystems are explained here:

Figure 6.6a corresponds to the dialogue box 'Three-Phase AC Data' in Figure 6.5. In this dialogue box, the RMS line-to-neutral voltage in volts and the frequency in hertz are entered. The peak voltage V_m and angular frequency ω are internally calculated. Figure 6.6b corresponds to the dialogue box 'Firing Angle' in Figure 6.5. The firing angle α in radians is entered in this dialogue box. Figure 6.6b has three constant blocks that generate α, $[(2\pi/3)+\alpha]$, $[(4\pi/3)+\alpha]$ for the three phases. Figure 6.6c generates the switch function SF for the three phases. The switch function corresponds to the firing pulses for the back-to-back connected thyristors in the three phases. These switch functions SF_A, SF_B and SF_C are generated by Relay, Relay1 and Relay2 in Figure 6.6c connected to input phases A, B and C, as defined in Equation 6.6:

Interactive Models for AC to AC Converters 125

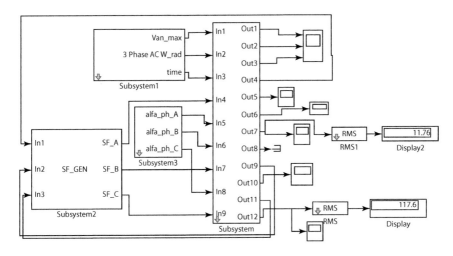

FIGURE 6.4
Model of three-phase fully controlled SCR controller with star-connected resistive load and isolated neutral for $\alpha = 30°$.

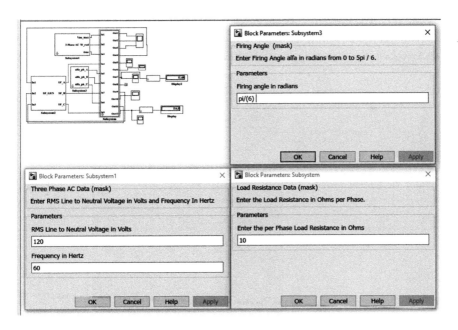

FIGURE 6.5
Model of three-phase fully controlled SCR controller with star-connected resistive load and isolated neutral for $\alpha = 30°$ with interactive blocks.

$$\left.\begin{array}{rl} SF\_A = +1 & \text{for} \quad \alpha \leq \omega.t \leq (\pi+\alpha) \\ = -1 & \text{for} \quad (\pi+\alpha) \leq \omega.t \leq (2\pi+\alpha) \\ SF\_B = +1 & \text{for} \quad \left(\frac{2\pi}{3}+\alpha\right) \leq \omega.t \leq \left(\frac{5\pi}{3}+\alpha\right) \\ = -1 & \text{for} \quad \left(\frac{5\pi}{3}+\alpha\right) \leq \omega.t \leq \left(\frac{8\pi}{3}+\alpha\right) \\ SF\_C = +1 & \text{for} \quad \left(\frac{4\pi}{3}+\alpha\right) \leq \omega.t \leq \left(\frac{7\pi}{3}+\alpha\right) \\ = -1 & \text{for} \quad \left(\frac{7\pi}{3}+\alpha\right) \leq \omega.t \leq \left(\frac{10\pi}{3}+\alpha\right) \end{array}\right\} \quad (6.6)$$

The model of the back-to-back connected thyristors in the three phases and the star-connected resistive load is shown in Figure 6.6d. The development of this model in Figure 6.6d is given as follows:

Referring to the topmost four-input mux corresponding to Phase A, the inputs $u(1)$ to $u(4)$ are, respectively, peak line-to-neutral voltage V_m ($\sqrt{2}.V_S$), supply angular frequency ω, time t and SCR firing angle α. The Fcn block at the top marked 'VAN' generates v_{AN} given by Equation 6.1, while the bottom Fcn block generates an attenuated value of v_{AN} lagging by the firing angle α. The two Fcn blocks corresponding to Phase B and C, respectively, generate

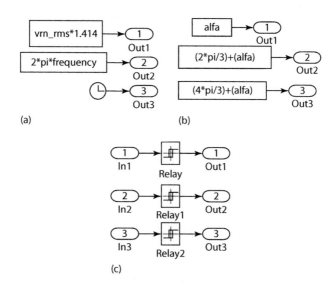

FIGURE 6.6
(a–d) Three-phase SCR controller with star-connected resistive load and isolated neutral–model subsystems.

Interactive Models for AC to AC Converters

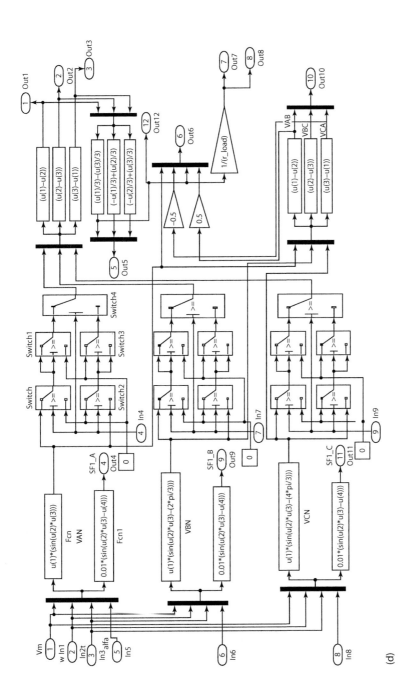

FIGURE 6.6
Continued

v_{BN} given by Equation 6.1 and attenuated v_{BN} lagging by $[(2.\pi/3)+\alpha]$, v_{CN} given by Equation 6.1 and attenuated v_{CN} lagging by $[(4.\pi/3)+\alpha]$. The bottom mux output of each phase marked SF1_A, SF1_B and SF1_C in Figure 6.6d are given as input In1, In2 and In3 to the three relays in Figure 6.6c. All three relays output 1 when the respective input crosses zero and becomes positive and output −1 when the respective input crosses zero and becomes negative.

Considering Phase A for clarity, SF1_A is passed as the $u(2)$ input to the threshold switches marked Switch and Switch2 in Figure 6.6d. The VAN output of the Fcn block is passed as the $u(1)$ input to Switch and the $u(3)$ input to Switch2. Zero is passed to the $u(3)$ input of Switch and the $u(1)$ input to Switch2. The output of Switch is passed to the $u(1)$ and $u(2)$ input of Switch1 and its $u(3)$ input is zero. The output of Switch2 is passed to the $u(2)$ and $u(3)$ input of Switch3 and its $u(1)$ input is zero. The output of Switch1 and Switch3 are passed as the $u(1)$ and $u(3)$ input to Switch4 and its $u(2)$ input is SF_A. Switch, Switch1, Switch2, Switch3 and Switch4 all have a threshold value of zero and all these switches' output correspond to $u(1)$ when $u(2)$ is greater than or equal to zero, else all these switches' output correspond to the $u(3)$ input. The threshold switches corresponding to Phase B and Phase C operate on the same principle as for Phase A.

Referring to the threshold switches in Phase A shown in Figure 6.6d and noting the definition of SF_A in Equation 6.6, the output of various threshold switches can be explained as follows.

Switch and Switch2 have the following output given by Equations 6.7 and 6.8, respectively:

$$\text{Switch output} = +V_m.\sin(\omega t) \quad \text{for} \quad \alpha \leq \omega t \leq (\pi+\alpha)$$
$$= 0 \quad \text{for} \quad \text{other time intervals} \tag{6.7}$$

$$\text{Switch2 output} = -V_m.\sin(\omega t) \quad \text{for} \quad (\pi+\alpha) \leq \omega t \leq (2\pi+\alpha)$$
$$= 0 \quad \text{for} \quad \text{other time intervals} \tag{6.8}$$

Switch1 and Switch3 have the following output given by Equations 6.9 and 6.10, respectively:

$$\text{Switch1 output} = +V_m.\sin(\omega t) \quad \text{for} \quad \alpha \leq \omega t \leq \pi$$
$$= 0 \quad \text{for} \quad \text{other time intervals} \tag{6.9}$$

$$\text{Switch3 output} = -V_m.\sin(\omega t) \quad \text{for} \quad (\pi+\alpha) \leq \omega t \leq 2\pi$$
$$= 0 \quad \text{for} \quad \text{other time intervals} \tag{6.10}$$

Switch4 has the following output given by Equation 6.11:

Interactive Models for AC to AC Converters

$$\text{Switch4 output} = +V_m.\sin(\omega t) \quad \text{for} \quad \alpha \le \omega t \le \pi$$

$$= -V_m.\sin(\omega t) \quad \text{for} \quad (\pi + \alpha) \le \omega t \le 2\pi \quad (6.11)$$

$$= 0 \quad \text{for} \quad \text{other time intervals}$$

The Switch4 output of Phase A and the corresponding Switch output of Phase B and Phase C are connected to a three-input mux as shown in Figure 6.6d. The three Fcn blocks calculate the three-phase line-to-line voltage across the star-connected resistive load using the following formula:

$$\begin{bmatrix} v_{L1L2} \\ v_{L2L3} \\ v_{L3L1} \end{bmatrix} = \begin{bmatrix} 1 & -1 & 0 \\ 0 & 1 & -1 \\ -1 & 0 & 1 \end{bmatrix} * \begin{bmatrix} v_{L1N} \\ v_{L2N} \\ v_{L3N} \end{bmatrix} = \begin{bmatrix} u(1) - u(2) \\ u(2) - u(3) \\ u(3) - u(1) \end{bmatrix} \quad (6.12)$$

In Equation 6.12, $L1$, $L2$, $L3$ and N correspond to the terminals A, B, C and N as marked in Figure 6.1. In Figure 6.6d, Out1, Out2 and Out3 correspond to these line-to-line voltages across the star-connected load. These three voltages are given to another set of three-input mux as shown in Figure 6.6d. The three Fcn blocks connected to this mux calculates the line-to-neutral voltages across the star-connected resistive load using the following formula:

$$\begin{bmatrix} v_{L1n} \\ v_{L2n} \\ v_{L3n} \end{bmatrix} = \begin{bmatrix} \dfrac{1}{3} & 0 & \dfrac{-1}{3} \\ \dfrac{-1}{3} & \dfrac{1}{3} & 0 \\ 0 & \dfrac{-1}{3} & \dfrac{1}{3} \end{bmatrix} * \begin{bmatrix} v_{L1L2} \\ v_{L2L3} \\ v_{L3L1} \end{bmatrix} = \begin{bmatrix} \dfrac{u(1)}{3} - \dfrac{u(3)}{3} \\ \dfrac{-u(1)}{3} + \dfrac{u(2)}{3} \\ \dfrac{-u(2)}{3} + \dfrac{u(3)}{3} \end{bmatrix} \quad (6.13)$$

In Equation 6.13, n is the load neutral point as shown in Figure 6.1. The line-to-neutral voltage v_{L1n} is multiplied by $1/(\text{r\_load})$ to obtain the load current through each of the resistors.

6.2.2 Simulation Results

The simulation of the fully controlled three-phase AC controller with star-connected load and neutral isolated was carried out using the ode23t (mod. stiff/trapezoidal) solver for various firing angles α of $\pi/6$, $\pi/4$, $\pi/3$, $\pi/2$ and $2.\pi/3$ rad [7]. The data used for this controller is given in Table 6.1 [1]. The simulation results for firing angles α of $\pi/6$, $\pi/4$, $\pi/3$, $\pi/2$ and $2.\pi/3$ rad are shown from Figures 6.7a–f to Figure 6.11a–f. The names of the various waveforms are given at the top of each simulation result. The simulation results for various firing angles α are tabulated in Table 6.2. The RMS line-to-neutral voltage

TABLE 6.1
Data for Three-Phase AC Controller with Star-Connected Resistive Load

Sl. No.	RMS Line-to-Neutral Voltage (V)	Frequency (Hz)	Firing Angle α (rad)	Per Phase Load Resistance (Ω)
1	120	60	$\pi/6$	10
2	120	60	$\pi/4$	10
3	120	60	$\pi/3$	10
4	120	60	$\pi/2$	10
5	120	60	$2.\pi/3$	10

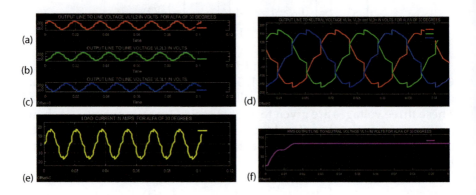

FIGURE 6.7
(a–f) Three-phase AC to AC controller with star-connected resistive load: simulation results for $\alpha = \pi/6$ rad.

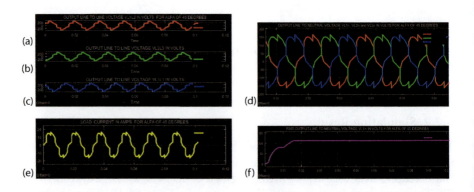

FIGURE 6.8
(a–f) Three-phase AC to AC controller with star-connected resistive load: simulation results for $\alpha = \pi/4$ rad.

Interactive Models for AC to AC Converters

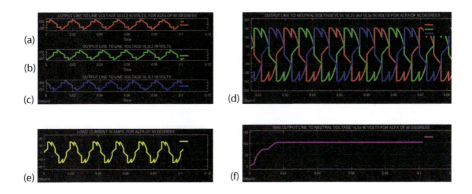

FIGURE 6.9
(a–f) Three-phase AC to AC controller with star-connected resistive load: simulation results for $\alpha = \pi/3$ rad.

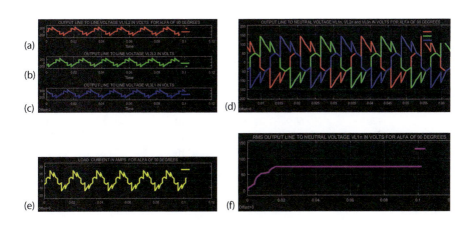

FIGURE 6.10
(a–f) Three-phase AC to AC controller with star-connected resistive load: simulation results for $\alpha = \pi/2$ rad.

across resistive load, calculated by using Equations 6.3 through 6.5, and also the RMS load current by calculation are given in Table 6.3.

Comparison of the values given in Tables 6.2 and 6.3 reveals that the percentage error for load current and RMS line-to-neutral output voltage is very small for $\alpha = \pi/6$, $\pi/4$ and $\pi/3$ rad and this value is in the range of 15.16% for $\alpha = \pi/2$ rad and -27.8% for $\alpha = 2\pi/3$ rad.

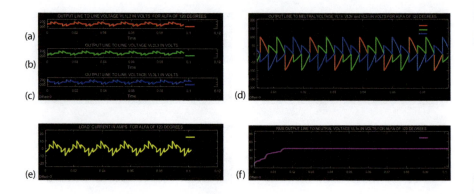

FIGURE 6.11
(a–f) Three-phase AC to AC controller with star-connected resistive load: simulation results for $\alpha = 2.\pi/3$ rad.

TABLE 6.2

Simulation Results for Three-Phase AC Controller with Star-Connected Resistive Load

Sl. No.	Line-to-Neutral Voltage (V)	Frequency (Hz)	Firing Angle α (rad)	Per Phase Load Resistance (Ω)	RMS Line-to-Neutral Load Voltage (V)	RMS Load Current (A)
1	120	60	$\pi/6$	10	117.6	11.76
2	120	60	$\pi/4$	10	112.5	11.25
3	120	60	$\pi/3$	10	103.2	10.32
4	120	60	$\pi/2$	10	74.82	7.482
5	120	60	$2.\pi/3$	10	43.32	4.332

TABLE 6.3

Calculated Values for Three-Phase AC Controller with Star-Connected Resistive Load

Sl. No.	Line-to-Neutral Voltage (V)	Frequency (Hz)	Firing Angle α (rad)	Per Phase Load Resistance (Ω)	RMS Line-to-Neutral Load Voltage (V)	RMS Load Current (A)
1	120	60	$\pi/6$	10	117.38	11.738
2	120	60	$\pi/4$	10	111.53	11.153
3	120	60	$\pi/3$	10	100.88	10.088
4	120	60	$\pi/2$	10	64.97	6.497
5	120	60	$2.\pi/3$	10	60	6.00

6.3 Analysis of a Fully Controlled Three-Phase AC Voltage Controller in Series with Resistive Load Connected in Delta

The three-phase three-wire AC voltage controller with back-to-back SCRs in series with resistive load connected in delta is shown in Figure 6.12 [1–4]. By varying the firing angle α, the RMS current through the load can be varied.

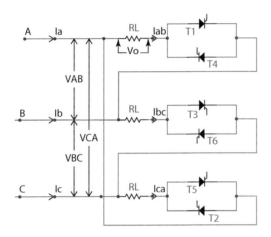

FIGURE 6.12
Three-phase thyristor AC controller in series with resistive load in delta.

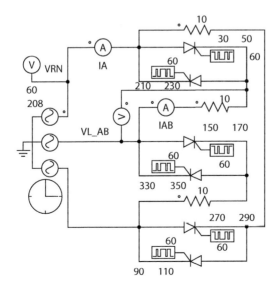

FIGURE 6.13
Three-phase AC controller in series with resistive load in delta.

The firing sequence is T1, T2, T3, T4, T5 and T6. The RMS output voltage is derived in Equation 6.15. The three-phase line-to-line AC voltage is defined in Equation 6.14.

$$\begin{aligned} v_{AB} &= \sqrt{2} * V_S * \sin(\omega t) \\ v_{BC} &= \sqrt{2} * V_S * \sin\left(\omega t - \frac{2\pi}{3}\right) \\ v_{CA} &= \sqrt{2} * V_S * \sin\left(\omega t + \frac{2\pi}{3}\right) \end{aligned} \quad (6.14)$$

In Equation 6.14, v_{AB}, v_{BC}, v_{CA} are the instantaneous voltages and V_S is the RMS line-to-line voltage. As the load resistance is in series with the back-to-back connected SCRs forming a delta configuration, the RMS phase voltage V_O can be derived as follows, for any firing angle α.

$$\begin{aligned} V_{O\_RMS} &= \sqrt{\frac{1}{\pi} * \int_{\alpha}^{\pi} 2V_S^2 * \sin^2(\omega t).d(\omega t)} \\ &= V_S * \sqrt{\left[1 - \frac{\alpha}{\pi} + \frac{\sin(2\alpha)}{2\pi}\right]} \end{aligned} \quad (6.15)$$

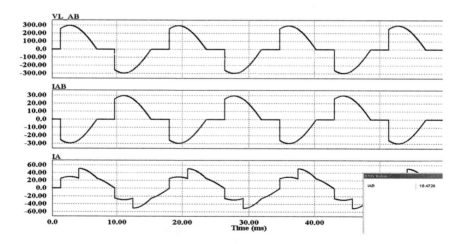

FIGURE 6.14
Three-phase AC controller in series with resistive load in delta: output waveforms for $\alpha = \pi/6$ rad.

Interactive Models for AC to AC Converters 135

The range of firing angles is $0 \leq \alpha \leq \pi$. In Figure 6.12, i_{ab}, i_{bc} and i_{ca} are the phase currents through the load. Also, $i_a = i_{ab} - i_{ca}$, $i_b = i_{bc} - i_{ab}$ and $i_c = i_{ca} - i_{bc}$ are the line currents.

A PSIM [5] circuit model of the three-phase AC to three-phase AC thyristor controller in series with resistive load connected in delta, for a firing angle of $\pi/6$ rad, is shown in Figure 6.13. The RMS line-to-line input voltage is 208 V and frequency is 60 Hz. The simulation results are shown in Figure 6.14. From simulation results, it is seen that the load current is 18.47 A for a firing angle of $\pi/6$ rad and the load voltage V_{AB} is 184.7 V with a load resistance of 10 Ω.

6.3.1 Modelling of a Fully Controlled Three-Phase AC Voltage Controller in Series with Resistive Load Connected in Delta

This section describes the modelling of the three-phase delta-connected AC controller shown in Figure 6.12 [6]. The model for a firing angle α of $\pi/6$ rad is shown in Figure 6.15. The various dialogue boxes are shown in Figure 6.16 and the various subsystems in Figure 6.17a–d. The various subsystems are explained here:

Figure 6.17a corresponds to the dialogue box 'Three-Phase AC Supply Data' in Figure 6.16. In this dialogue box, the RMS line-to-line voltage in volts and the frequency in hertz are entered. The peak voltage V_m and angular frequency ω are internally calculated. Figure 6.17b corresponds to the dialogue box 'Firing Pulse Generator' in Figure 6.16. The firing angle α in radians is entered in this dialogue box. Figure 6.17b has three constant blocks that generate α, $[(2\pi/3) + \alpha]$ and $[(4\pi/3) + \alpha]$ for the three phases. Figure 6.17c generates the switch function SF for the three phases. The switch function corresponds to the firing pulses for the back-to-back connected thyristors in the three phases. These switch functions SF_A, SF_B and SF_C are defined as follows:

The dialogue box 'Load Resistance' in Figure 6.16 corresponds to Figure 6.17d. The load resistance value in ohms per phase is entered in this dialogue box. Figure 6.17d corresponds to the model of the back-to-back connected thyristors in the three phases in series with load resistors connected in delta. The development of this model in Figure 6.17d is given here:

Referring to the topmost four-input mux corresponding to Phase A, the inputs $u(1)$ to $u(4)$ are, respectively, peak line-to-line voltage V_m ($\sqrt{2}.V_S$), supply angular frequency ω, time t and firing angle α. The function block at the top marked Fcn generates v_{AB} given by Equation 6.14, while the bottom Fcn1 block generates attenuated v_{AB} lagging v_{AB} by the firing angle α. The two Fcn blocks corresponding to Phase B and C, respectively, generate v_{BC} given by Equation 6.14 and attenuated v_{BC} lagging v_{AB} by $[(2.\pi/3) + \alpha]$, v_{CA} given by Equation 6.14 and attenuated v_{CA} lagging v_{AB} by $[(4.\pi/3) + \alpha]$.

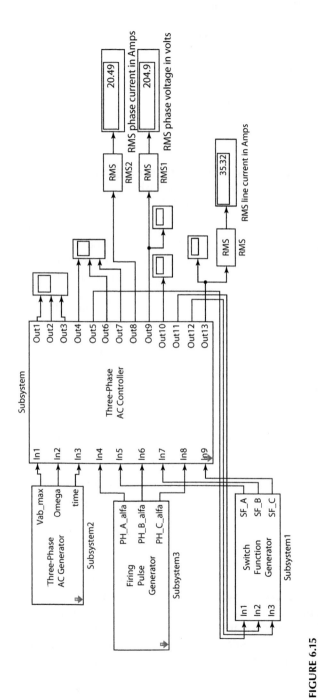

FIGURE 6.15
Three-phase AC controller: back-to-back thyristor in series with resistive load connected in delta.

Interactive Models for AC to AC Converters 137

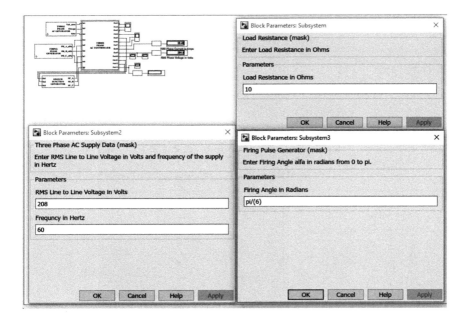

FIGURE 6.16
Three-phase AC controller: back-to-back thyristor in series with resistive load connected in delta with interactive blocks.

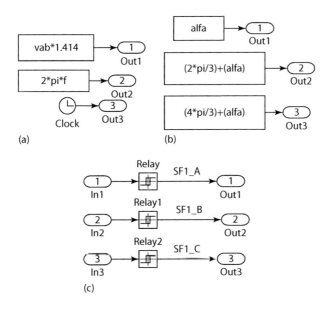

FIGURE 6.17
(a–d) Three-phase AC controller: back-to-back thyristor in series with resistive load connected in delta–model subsystems.

138 Power Electronic Converters

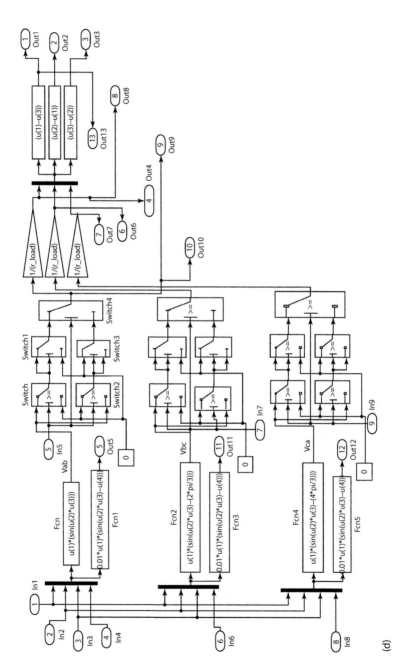

FIGURE 6.17
Continued

$$SF\_A = +1 \quad \text{for} \quad \alpha \le \omega.t \le (\pi + \alpha)$$

$$= -1 \quad \text{for} \quad (\pi + \alpha) \le \omega.t \le (2\pi + \alpha)$$

$$SF\_B = +1 \quad \text{for} \quad \left(\frac{2\pi}{3} + \alpha\right) \le \omega.t \le \left(\frac{5\pi}{3} + \alpha\right)$$

$$= -1 \quad \text{for} \quad \left(\frac{5\pi}{3} + \alpha\right) \le \omega.t \le \left(\frac{8\pi}{3} + \alpha\right) \tag{6.16}$$

$$SF\_C = +1 \quad \text{for} \quad \left(\frac{4\pi}{3} + \alpha\right) \le \omega.t \le \left(\frac{7\pi}{3} + \alpha\right)$$

$$= -1 \quad \text{for} \quad \left(\frac{7\pi}{3} + \alpha\right) \le \omega.t \le \left(\frac{10\pi}{3} + \alpha\right)$$

The bottom function block output, that is, the output of Fcn1, Fcn3 and Fcn5 of each phase marked in Figure 6.17d are passed as input In1, In2 and In3 to the three relays in Figure 6.17c. All three relays output 1 when the respective input crosses zero and becomes positive, and output −1 when the respective input crosses zero and becomes negative.

Considering Phase A for clarity, SF_A is passed as the $u(2)$ input to the threshold switches marked Switch and Switch2 in Figure 6.17d. The output of the Fcn block is passed as the $u(1)$ input to Switch and the $u(3)$ input to Switch2. Zero is passed to the $u(3)$ input of Switch and the $u(1)$ input of Switch2. The output of Switch is passed to the $u(1)$ and $u(2)$ input of Switch1 and its $u(3)$ input is zero. The output of Switch2 is passed to the $u(2)$ and $u(3)$ input of Switch3 and its $u(1)$ input is zero. The output of Switch1 and Switch3 are passed as the $u(1)$ and $u(3)$ input to Switch4 and its $u(2)$ input is SF_A. Switch, Switch1, Switch2, Switch3 and Switch4 all have a threshold value of zero and all these switches' output correspond to $u(1)$ when $u(2)$ is greater than or equal to zero, else all these switches' output correspond to the $u(3)$ input. The threshold switches corresponding to Phase B and Phase C operate on the same principle as for Phase A.

Referring to the threshold switches in Phase A shown in Figure 6.17d and noting the definition of SF_A in Equation 6.16, the output of various threshold switches can be explained as follows:

Switch and Switch2 have the following output, given by Equation 6.17 and 6.18, respectively.

$$\text{Switch output} = +V_m.\sin(\omega t) \quad \text{for} \quad \alpha \le \omega t \le (\pi + \alpha)$$

$$= 0 \quad \text{for} \quad \text{other time intervals} \tag{6.17}$$

$$\text{Switch2 output} = -V_m.\sin(\omega t) \quad \text{for} \quad (\pi + \alpha) \le \omega t \le (2\pi + \alpha)$$

$$= 0 \quad \text{for} \quad \text{other time intervals} \tag{6.18}$$

Switch1 and Switch3 have the following output, given by Equations 6.19 and 6.20, respectively:

$$\text{Switch1 output} = +V_m.\sin(\omega t) \quad \text{for} \quad \alpha \le \omega t \le \pi$$
$$= 0 \quad \text{for} \quad \text{other time intervals} \tag{6.19}$$

$$\text{Switch3 output} = -V_m.\sin(\omega t) \quad \text{for} \quad (\pi + \alpha) \le \omega t \le 2\pi$$
$$= 0 \quad \text{for} \quad \text{other time intervals} \tag{6.20}$$

Switch4 has the following output, given by Equation 6.21:

$$\text{Switch4 output} = +V_m.\sin(\omega t) \quad \text{for} \quad \alpha \le \omega t \le \pi$$
$$= -V_m.\sin(\omega t) \quad \text{for} \quad (\pi + \alpha) \le \omega t \le 2\pi \tag{6.21}$$
$$= 0 \quad \text{for} \quad \text{other time intervals}$$

The Switch4 output in Figure 6.17d corresponds to the phase voltage V_O across the load resistor. This V_O is multiplied by 1/(r_load) using gain multiplier blocks to obtain the phase currents i_{ab}, i_{bc} and i_{ca} through the load resistor. These respective phase currents are passed to a three-input mux. The three Fcn blocks performing $(u(1) - u(3))$, $(u(2) - u(1))$ and $(u(3) - u(2))$ output the line currents i_a, i_b and i_c.

6.3.2 Simulation Results

Simulation of the fully controlled three-phase AC controller in series with resistive load connected in delta was carried out using the ode23t (mod. stiff/trapezoidal) solver for various firing angles α of $\pi/6$, $\pi/4$, $\pi/3$, $\pi/2$, $2.\pi/3$ and π rad [7]. The data used for this controller are given in Table 6.4 [1]. The simulation results for firing angles α of $\pi/6$, $\pi/4$, $\pi/3$, $\pi/2$, $2.\pi/3$ and π rad are shown from Figure 6.18a–h to Figure 6.23a–h, respectively.

TABLE 6.4

Data for Three-Phase AC Controller with Delta-Connected Resistive Load

Sl. No.	RMS Line-to-Line Supply Voltage (V)	Frequency Hz	Firing Angle α (rad)	Per Phase Load Resistance (Ω)
1	208	60	$\pi/6$	10
2	208	60	$\pi/4$	10
3	208	60	$\pi/3$	10
4	208	60	$\pi/2$	10
5	208	60	$2.\pi/3$	10
6	208	60	π	10

Interactive Models for AC to AC Converters 141

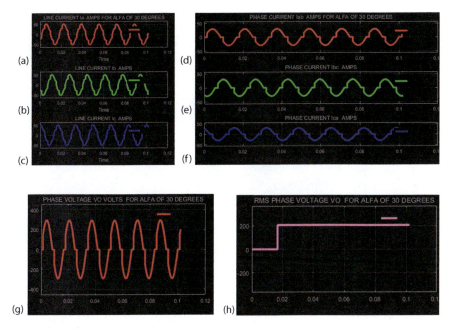

FIGURE 6.18
(a–h) Three-phase AC controller with resistive load in delta: simulation results for $\alpha = \pi/6$ rad.

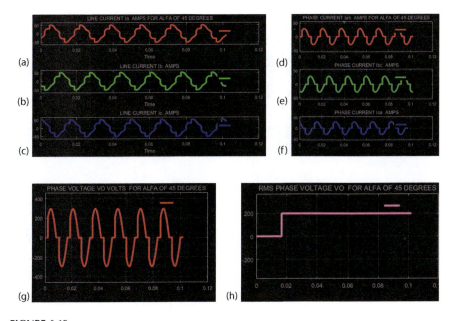

FIGURE 6.19
(a–h) Three-phase AC controller with resistive load in delta: simulation results for $\alpha = \pi/4$ rad.

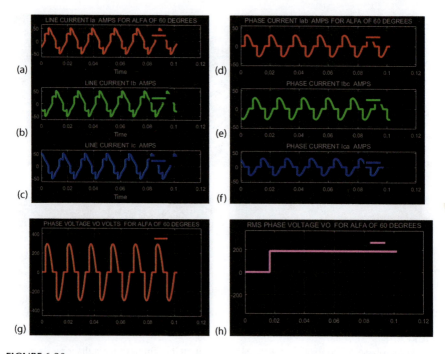

FIGURE 6.20
(a–h) Three-phase AC controller with resistive load in delta: simulation results for $\alpha = \pi/3$ rad.

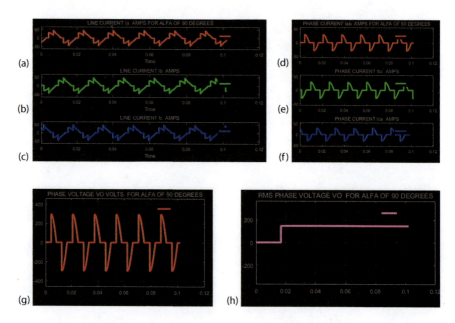

FIGURE 6.21
(a–h) Three-phase AC controller with resistive load in delta: simulation results for $\alpha = \pi/2$ rad.

Interactive Models for AC to AC Converters 143

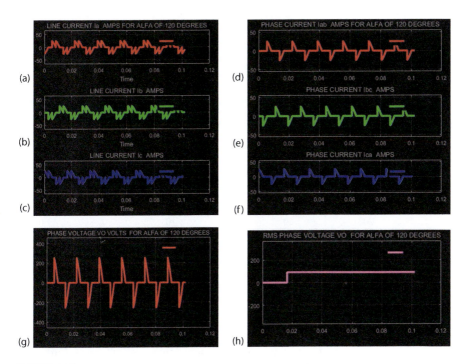

FIGURE 6.22
(a–h) Three-phase AC controller with resistive load in delta: simulation results for $\alpha = 2\pi/3$ rad.

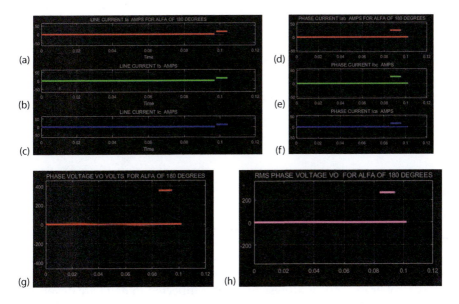

FIGURE 6.23
(a–h) Three-phase AC controller with resistive load in delta: simulation results for $\alpha = \pi$ rad.

TABLE 6.5

Simulation Results for Three-Phase AC Controller in Series with Resistive Load Connected in Delta

Sl. No.	RMS Line-to-Line Input Voltage (V)	Frequency (Hz)	Firing Angle α (rad)	Per Phase Load Resistance (Ω)	RMS Load Voltage V_O (V)	RMS Load Current i_{ab} (A)
1	208	60	$\pi/6$	10	204.9	20.49
2	208	60	$\pi/4$	10	198.3	19.83
3	208	60	$\pi/3$	10	186.5	18.65
4	208	60	$\pi/2$	10	147.1	14.71
5	208	60	$2.\pi/3$	10	91.96	9.196
6	208	60	π	10	0	0

TABLE 6.6

Calculated Values for Three-Phase AC Controller in Series with Resistive Load Connected in Delta

Sl. No.	RMS Line-to-Line Input Voltage (V)	Frequency (Hz)	Firing Angle α (rad)	Per Phase Load Resistance (Ω)	RMS Load Voltage V_O (V)	RMS Load Current i_{ab} (A)
1	208	60	$\pi/6$	10	204.978	20.497
2	208	60	$\pi/4$	10	198.327	19.832
3	208	60	$\pi/3$	10	186.56	18.65
4	208	60	$\pi/2$	10	147.078	14.7078
5	208	60	$2.\pi/3$	10	91.968	9.1968
6	208	60	π	10	0	0

The simulation results and the calculated values for various firing angles α for the three-phase back-to-back connected thyristor AC controller with series-connected resistive load in delta are shown in Tables 6.5 and 6.6, respectively [6].

6.4 Conclusions

System models for three-phase back-to-back connected thyristor AC controllers in series with lines connected to star-connected resistive load with isolated neutral, and also for three-phase back-to-back connected thyristor AC controller in series with resistive load connected in delta are developed, and their simulation and theoretically calculated results are presented. The switching function concept is used in these two models. It is seen from

Tables 6.2 and 6.3 that as the firing angle α increases, the percentage error between the theoretical and simulated results for line-to-neutral voltage across the load resistor increases [6]. Comparison of the values given in Tables 6.2 and 6.3 reveals that the percentage error for load current and RMS line-to-neutral output voltage is very small for $\alpha = \pi/6$, $\pi/4$ and $\pi/3$ rad and this value is in the range of 15.16% for $\alpha = \pi/2$ rad and -27.8% for $\alpha = 2\pi/3$ rad. It is also seen from Tables 6.5 and 6.6 that for the various firing angles α, the percentage error between the theoretical and simulated values for output voltage across load and load current per phase is negligible [6].

References

1. M.H. Rashid: *Power Electronics Circuits, Devices and Applications*, Upper Saddle River, NJ: Pearson Education, Pearson Prentice Hall, 2004.
2. M.H. Rashid [Ed.]: *Power Electronics Handbook*, Burlington, MA: Elsevier, 2011.
3. D.W. Hart: *Introduction to Power Electronics*, Upper Saddle River, NJ: Prentice Hall, 1997.
4. A.M. Trzynadlowski: *Introduction to Modern Power Electronics*, Hoboken, NJ: John Wiley, 2010; pp. 199–209.
5. Powersim Inc.: "PSIM demo version 10.0", 2015.
6. N.P.R. Iyer: "MATLAB/SIMULINK modules for modelling and simulation of power electronic converters and electric drives", M.E. by research thesis, University of Technology Sydney, NSW, Australia, 2006, Chapter 6.
7. The Mathworks Inc.: "MATLAB/Simulink" Release notes, version 8.8, R2016b, 2016.

7

Interactive Modelling of an Switched Mode Power Supply Using Buck Converter

7.1 Introduction

Switched mode power supplies (SMPS) are used to obtain a regulated DC output voltage from an unregulated DC input voltage. Here, the output voltage V_O is compared with a reference voltage V_{REF} and the difference voltage, called *error voltage* (V_{ERR}), is compared with a high-frequency triangle carrier signal V_c to produce gate pulses to drive the buck converter semiconductor switch. This semiconductor switch may be a BJT, MOSFET or IGBT. Depending on the value of V_O, the duty cycle of the pulse width modulation (PWM) switching pulses adjust to maintain V_O almost nearly constant, irrespective of variations in the input DC voltage V_s. In this type of SMPS, the buck converter is always operated in the continuous conduction mode (CCM). The switching function concept is used to model the buck converter SMPS.

7.2 Principle of Operation of Switched Mode Power Supply

Figure 7.1 gives the general block diagram of the SMPS using a buck converter. Referring to Figure 7.1, the buck converter output V_O is attenuated using a potential divider and the resulting output V_{POT} is subtracted from the reference input V_{REF} using a difference amplifier to obtain the error output V_{ERR}. This V_{ERR} output is compared with triangle carrier V_c in an op-amp comparator to obtain the PWM switching pulses V_g to drive the buck converter semiconductor switch. The switch SS1 is used to select the direct mode or PI controller mode. The PI controller in Figure 7.1 is used to reduce the peak value of oscillations and its number of cycles in the output voltage V_O. Depending on the value of V_O, the duty cycle of the PWM switching pulses adjusts to maintain the output voltage V_O almost nearly constant, irrespective of variations in the input DC voltage V_S within a given range [1–6].

147

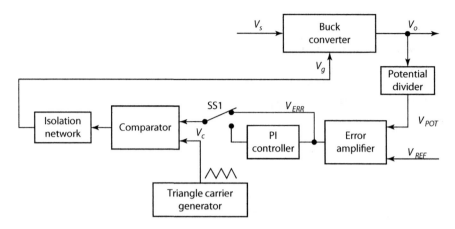

FIGURE 7.1
General model of the buck converter SMPS.

Simulation of SMPS using a buck converter was carried out using PSIM [7]. The parameters of the SMPS used for simulation are shown in Table 7.1.

The buck converter circuit simulation schematics using PSIM 6.1, without and with a PI controller, are shown in Figure 7.2a, b, respectively. The Gain block K with value −1 shown in Figure 7.2a, b is the inverting op-amp (a 10K resistor between the inverting pin and the PI controller; another 10K resistor between the output and the inverting pin and the non-inverting pin is earthed, as shown in Figure 7.2c). This inverting op-amp configuration is not shown in Figure 7.2a, b due to node limitation in the demo version of PSIM 6.1. The potential divider output V_{POT} is eight-ninths of the output voltage V_O. The reference voltage V_{REF} is 9 V. The error amplifier generates V_{ERR}, which is $(V_{REF} - V_{POT})$. The PI controller generates the transfer function $(0.1 + (6/s))$. The triangle carrier generator generates triangle carrier signal $V\_CARRIER$ with a frequency of 100 kHZ. The $V\_CARRIER$ output and V_{ERR} output are compared in an op-amp comparator. The output V_DRIVE is the gate pulse drive to the buck converter switch.

TABLE 7.1

Parameters of SMPS

Sl.No.	Parameter	Value	Units
1	Normal input voltage V_s	15	Volts
2	Normal output voltage V_O	9	Volts
3	Variation of input voltage	12–18	Volts
4	Load resistance R_L	10	Ohms
5	Inductance L_1	300E−6	Henries
6	Filter capacitance C_F	200E−6	Farads
7	Switching frequency F_{sw}	100	Kilohertz
8	PI controller – K_p, K_i	0.1,6	

Interactive Modelling of an SMPS Using Buck Converter 149

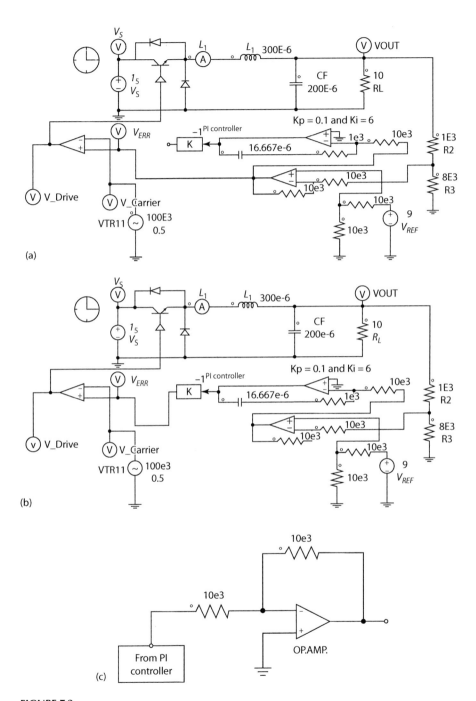

FIGURE 7.2
Buck converter: (a) SMPS without PI controller. (b) SMPS with PI controller. (c) Inverting operational amplifier.

150   *Power Electronic Converters*

The simulation of this buck converter SMPS with and without a PI controller was carried out using the demo version of PSIM 6.1. The simulation results without a PI controller and with a PI controller are shown in Figures 7.3 and 7.4, respectively.

7.3 Modelling of the Switched Mode Power Supply

The interactive model of the SMPS is shown in Figure 7.5 [8, 9]. The various dialogue boxes and help file are shown in Figures 7.6 and 7.7. The various subsystems are shown in Figure 7.8a–d. A brief description of the various subsystems is given as follows:

Figure 7.8a, b shows the model of the buck converter used in the SMPS. This corresponds to the dialogue box 'Buck Converter Unit' shown in Figure 7.6. The parameters relating to inductance, load resistance and filter capacitance are entered in the dialogue box. The buck converter circuit is shown in Figure 5.1. The modelling of the buck converter using the switching function concept has already been discussed in Section 5.4. The same modelling Equations (5.13 through 5.16) are used to model the buck converter used in this SMPS. As this SMPS is always operated in CCM, the inductor current switch function SF_IL1 has a value of unity in Equation 5.16. Figure 7.8a calculates di_{L1}/dt using Equation 5.16 with SF_IL1 replaced with unity and then integrated using the Integrator block to obtain i_{L1}. In Figure 7.8b, this inductor current i_{L1} is passed as input to a transfer function block to obtain the output voltage V_O. This transfer function block is the parallel combination of load resistor R_L and filter capacitor C_F. Figure 7.8c corresponds to the dialogue box 'Error Amplifier and PI Controller Unit' in Figure 7.6. The potential divider constant, reference voltage, proportional constant Kp and integral constant Ki are entered in the appropriate places in this dialogue box. The output voltage V_O is multiplied by this potential divider constant k_{pot} and compared in the error detector with the reference voltage V_{REF}, which produces the error output voltage V_{ERR}. A selector switch SS1 selects V_{ERR} directly or through the PI controller. This output from SS1 is passed to the comparator block in Figure 7.8d. Figure 7.8d corresponds to the dialogue box 'Triangle Carrier Generator Unit' in Figure 7.6 and the help file shown in Figure 7.7. Here, the triangle carrier pulse is generated by integration of the square pulse. The amplitude of this square pulse, the switching period of SMPS, the gain constant $k1$, and the two function constants $k2$ and c are entered in the appropriate places in the dialogue box. In Figure 7.8d, the Fcn2 block subtracts the amplitude V_{max} of the square pulse from half its value there by generating a square pulse with positive and negative peaks of $+V_{max}/2$ and $-V_{max}/2$, respectively, with a switching period of the SMPS Tsw. This is multiplied by $k1$ using the Gain2 block; then, this value is subtracted from a constant

Interactive Modelling of an SMPS Using Buck Converter   151

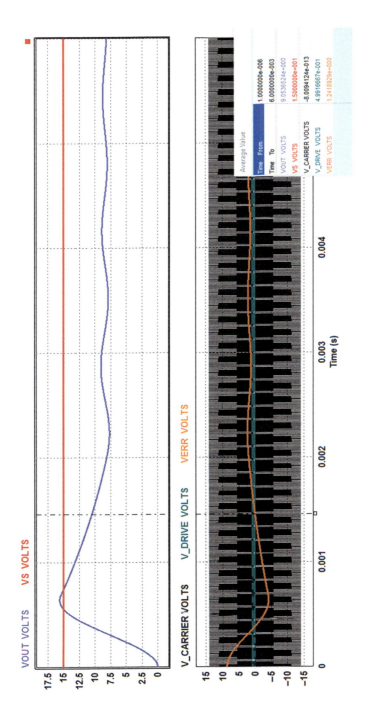

FIGURE 7.3
Buck SMPS: simulation results without PI controller.

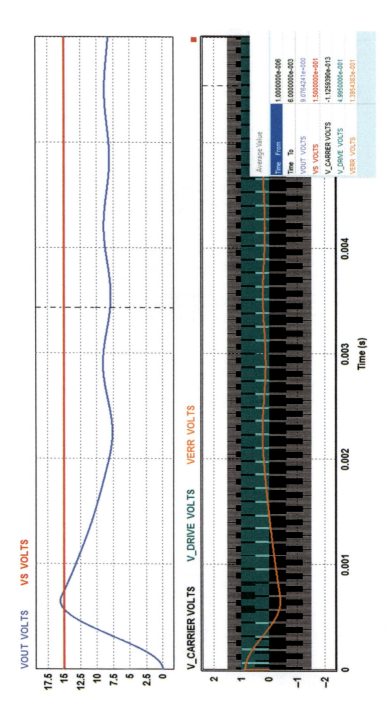

FIGURE 7.4
Buck SMPS: simulation results with PI controller.

Interactive Modelling of an SMPS Using Buck Converter

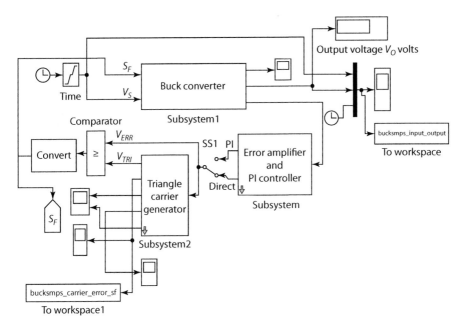

FIGURE 7.5
Buck converter SMPS.

c and the result is multiplied by another constant $k2$ using the Fcn3 block to obtain the triangle carrier wave with the switching period of the SMPS. The dialogue box 'Triangle Carrier Generator Unit' in Figure 7.7 provides the method to find constants $k1$, $k2$ and c and the peak value of the triangle carrier wave for any carrier switching frequency of the SMPS. The triangle carrier output V_{TRI} and the error output V_{ERR} either directly or through the PI controller are compared in the relational operator block, which is the comparator unit. The convert block converts the logical or Boolean output of the comparator to unsigned integer output. The comparator and convert blocks are shown in Figure 7.5. The selection of the relational operator for the comparator is also explained in Figure 7.7. The output of the comparator is the switch function SF, which drives the gate of the buck converter switch.

7.3.1 Simulation Results

Simulation of the buck converter SMPS was conducted with and without a PI controller using the ode23t(mod. stiff/trapezoidal) solver [10]. The parameters of the SMPS are as shown in Table 7.1. A time-varying input DC voltage source using a look-up table was used in both cases. The simulation results of the input voltage, output voltage, triangular carrier, error voltage output V_{ERR}, switch function SF and the inductor current IL without the PI controller are shown in Figure 7.9a–f. The same results with the PI controller are shown in Figure 7.10a–f, respectively.

154 Power Electronic Converters

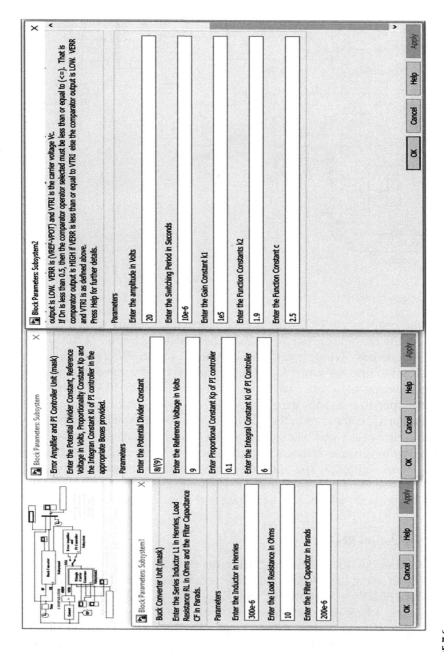

FIGURE 7.6
Buck converter SMPS with interactive blocks.

DETERMINATION OF PEAK VALUE OF TRIANGLE CARRIER WAVE:
To determine the peak value of triangle carrier wave, follow the procedure given below:
Let Vref be the reference voltage, Vpot be the potential divider output in the feedback path and Vpeak be the peak value traingle carrier wave. Then follow steps in square brackets.
[1] Calculate Verr = Vref - Vpot.
[2] Calculate Ton = Dn * Tsw.
If Ton is greater than or equal to Tsw / 2, then follow steps 3 to 6.
[3] Calculate Toff = Tsw - Ton.
[4] Find T1 = {(Tsw /2) - Toff } /2.
[5] Find slope of Triangle carrier wave m_tri = { Verr } / T1.
[6] Calculate Vpeak = m_tri * Tsw /4.
If Ton is less than Tsw /2, then follow steps 7 and 8.
[7] Find T1 = {(Tsw /2) - Ton } /2.
[8] Then follow steps 5 and 6 given above.
SELECTION OF CONSTANTS k1, k2 AND c:
The value of k1 is selcted consistant with the entered values of Vsqm and Tsw. If Tsw is an integer in microseconds then k1 will be a constant multiple of 1e5.
The value of function constant c will be [Vm(triangle) /2].
The value of k2 is suitably adjusted to get normal output voltage for normal input voltage without any type of PI or other controllers.
NOTE:
This triangle carrier generator is connected to the comparator block. If Vsn and Voutn are the normal input and normal output voltage respectively, then the normal duty cycle Dn of the buck converter is
Dn = Voutn / Vsn. Then slect the comparator operator as follows:
If Dn is equal to or greater than 0.5, then comparator operator selected must be greater than or equal to (>=). That is comparator output is HIGH if VERR is greater than or equal to VTRI else the comparator
output is LOW. VERR is (VREF-VPOT) and VTRI is the carrier voltage Vc.
If Dn is less than 0.5, then the comparator operator selected must be less than or equal to (<=). That is comparator output is HIGH if VERR is less than or equal to VTRI else the comparator output is LOW.
VERR and VTRI is as defined above.
Press Help for further details.

FIGURE 7.7
Buck SMPS: triangle carrier generator unit.

7.4 Conclusions

Examination of the simulation results of the buck SMPS model in Figure 7.9a–f and in Figure 7.10a–f indicates the following points:

- Without the PI controller, the number of initial switching cycles is as high as 11 for the first 0.01 s.
- With the PI controller, the number of initial switching cycles is reduced from 11 to 7 for the first 0.01 s.
- The peak-to-peak voltage output during the initial or first switching cycle is reduced by around 2.4 V with the PI controller as compared with without the PI controller.
- The duty cycle of SF adjusts to a suitable value to maintain the output voltage V_O close to 9 V.
- The change in output voltage for a given change in input voltage or regulation of the SMPS is greater with the PI controller as compared with without the PI controller.
- The normal output voltage is close to 9 V for a normal input voltage of 15 V in both cases.
- The triangle carrier peak with the PI controller is less than that without the PI controller to achieve the same DC output voltage of 9 V with a DC input voltage of 15 V in both cases.

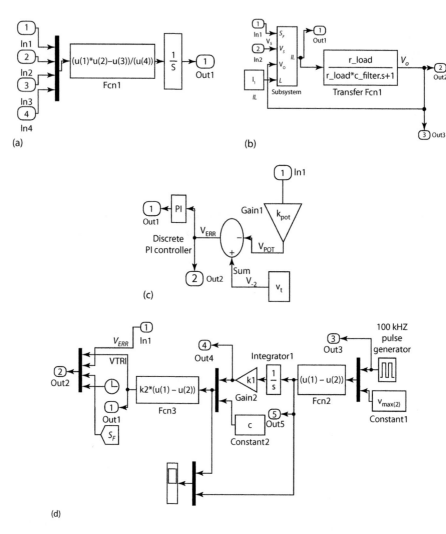

FIGURE 7.8
(a–d) Buck converter SMPS model subsystems.

- The inductor current is continuous and the buck converter operates in CCM in both cases.
- Depending on the chosen values of the parameters for the buck converter, the number of oscillation cycles and their peak values will change in both cases.

To conclude, the buck SMPS with PI controller reduces the number and peak-to-peak value of oscillation cycles in the output voltage as compared with the one without PI controller. However, the problem of the buck SMPS with PI controller is that the change in output voltage for a given change in

Interactive Modelling of an SMPS Using Buck Converter 157

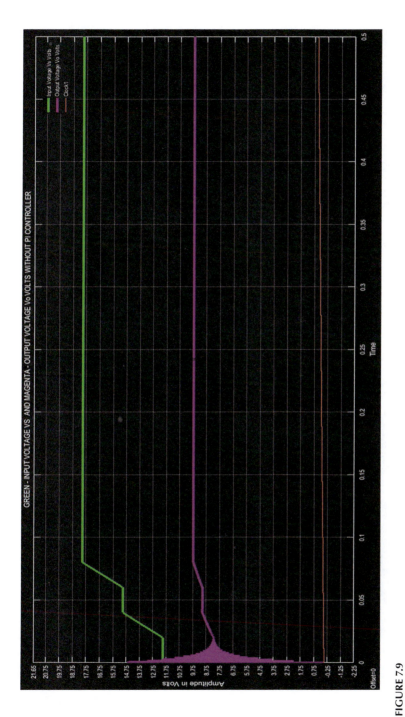

FIGURE 7.9
(a, b) Input voltage (green) vs. output voltage (magenta) without pi controller.

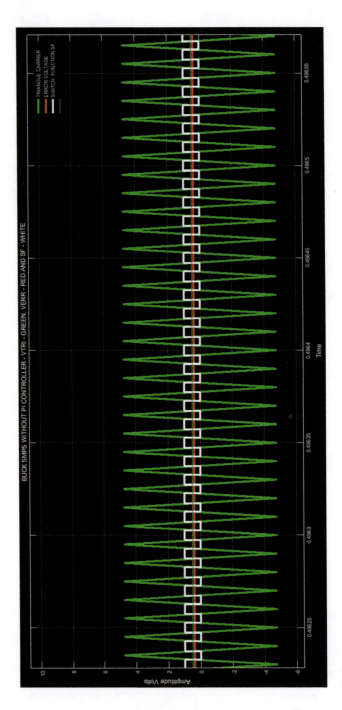

FIGURE 7.9
(c–e) Buck SMPS without pi controller. VTRI (green), VERR (red) and SF (white).

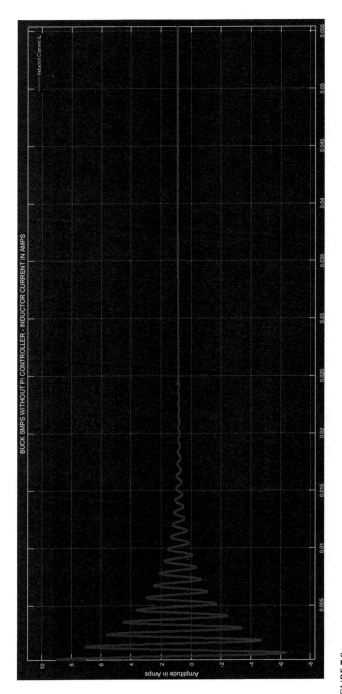

FIGURE 7.9
(f) Buck SMPS without pi controller: inductor current in amps.

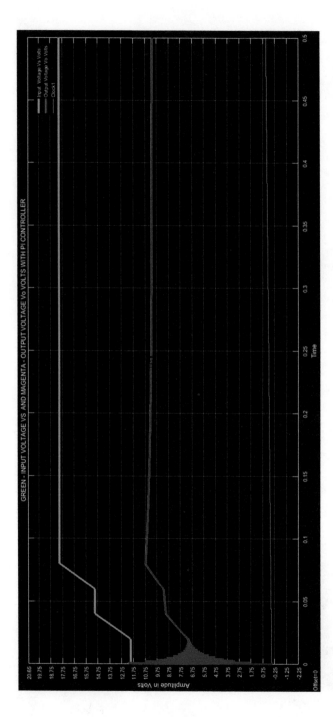

FIGURE 7.10
(a, b) Input voltage (green) vs. output voltage (magenta) with pi controller.

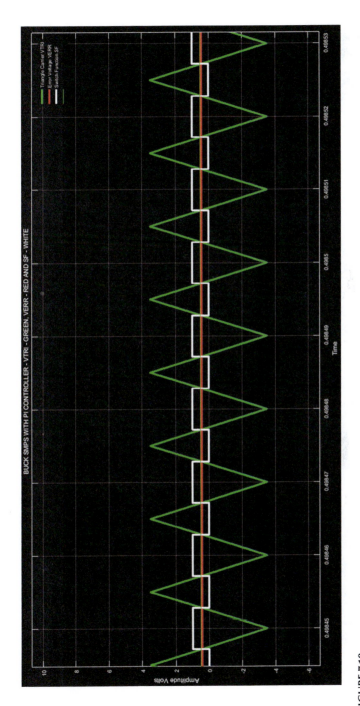

FIGURE 7.10
(c–e) Buck SMPS with pi controller. VTRI (green), VERR (red) and SF (white).

162 *Power Electronic Converters*

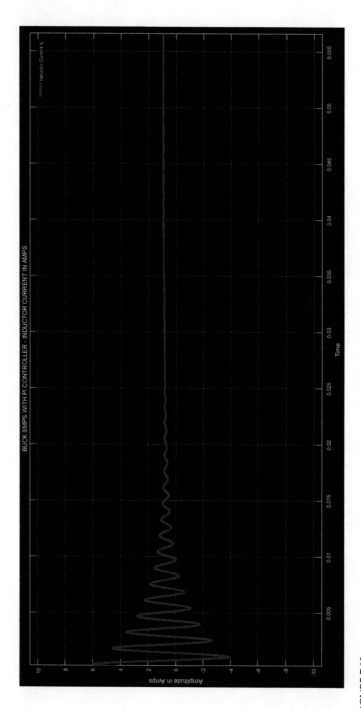

FIGURE 7.10
(f) Buck SMPS with pi controller: inductor current in amps.

input voltage is much larger compared with the one without PI controller. Referring to Figure 7.9a,b, it is seen that for the buck SMPS without PI controller, when the input voltage changes from 12 to 18 V, the steady-state output voltage changes from 8.25 to 9.75 V. From Figure 7.10a,b, it is seen that for the buck SMPS with PI controller, this steady-state output voltage variation is 7–10.25 V for the same change in input voltage.

References

1. M.H. Rashid: *Power Electronics Circuits, Devices and Applications*, Upper Saddle River, NJ: Pearson Education, Pearson Prentice Hall, 2004.
2. I. Batarseh: *Power Electronic Circuits*, Hoboken, NJ: Wiley, 2004.
3. D.W. Hart: *Introduction to Power Electronics*, Upper Saddle River, NJ: Prentice Hall, 1997.
4. N. Mohan, T.M. Undeland, and W.P. Robbins: *Power Electronics: Converters, Applications and Design*, Hoboken, NJ: Wiley, 1995.
5. B. Choi: *Pulse Width Modulated DC to DC Power Conversion: Circuit, Dynamics and Control Designs*, Piscataway, NJ: IEEE Press – Wiley, 2013, pp. 93–143.
6. F.L. Luo and H. Ye: *Power Electronics*, Boca Raton, FL: CRC Press, 2013.
7. Powersimtech.com: PSIM demo version, 2016.
8. N.P.R. Iyer and V. Ramaswamy "Modeling and simulation of a switched mode power supply using SIMULINK", *Australasian Universities Power Engineering Conference (AUPEC)*, Hobart, Tasmania, September 2005, pp. 562–567.
9. N.P.R. Iyer: "MATLAB/SIMULINK modules for modelling and simulation of power electronic converters and electric drives", M.E. by research thesis, University of Technology Sydney, NSW, Australia, 2006, Chapter 10.
10. The Mathworks Inc.: "MATLAB/Simulink Release Notes", R2016b, 2016.

8

Interactive Models for Fourth-Order DC to DC Converters

8.1 Introduction

As already mentioned in Chapter 5, fourth-order DC to DC converters have two inductors, one energy transfer capacitor, one filter capacitor, one or more diodes and a load resistor, along with the semiconductor switch, gate drive and DC voltage source. These converters can be operated in the buck or boost mode by proper choice of duty cycle for the gate pulse drive. Interactive system models for selected fourth-order DC to DC converters such as the single-ended primary inductance converter (SEPIC), quadratic boost and ultra-lift Luo converters are presented in this chapter for both continuous conduction mode (CCM) and discontinuous conduction mode (DCM).

8.2 Analysis of SEPIC Converter in CCM

The SEPIC converter is used to obtain a positive output voltage less than, equal to or greater than the input DC voltage. The circuit schematic of the SEPIC converter is shown in Figure 8.1 [1–5].

The switch is driven by a gate pulse whose period is T and duty cycle D. When the switch S_1 is ON for the interval $0 \leq t \leq D.T$ s, the diode D_1 is reverse biased and is open. When switch S_1 is OFF for the interval $D.T \leq t \leq T$ s, the diode D_1 is forward biased and provides a short circuit path. This is shown in Figure 8.2a, b. The average inductor current of L_1 and L_2 over one switching cycle is zero. This can be expressed as follows:

$$\frac{V_S.D.T}{L_1} + \frac{(V_S - V_{C1} - V_O).(1-D).T}{L_1} = 0 \tag{8.1}$$

165

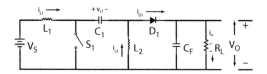

FIGURE 8.1
SEPIC converter.

$$\frac{(-V_{C1}).D.T}{L_2} + \frac{(V_O).(1-D).T}{L_2} = 0 \quad (8.2)$$

From Equation 8.2, V_{C1} can be expressed as follows:

$$V_{C1} = \frac{V_O.(1-D)}{D} \quad (8.3)$$

Using Equation 8.3 in Equation 8.1 and simplifying gives the following result:

$$\frac{V_O}{V_S} = \frac{D}{(1-D)} \quad (8.4)$$

The polarity of V_O is positive at the top and negative at the bottom terminal, as shown in Figure 8.1. The relevant waveforms for the SEPIC converter CCM operation are shown in Figure 8.3. Neglecting switching losses, the input power and output power are equal. This can be expressed as follows:

$$V_S * I_{L1} = \frac{V_O^2}{R_L} \quad (8.5)$$

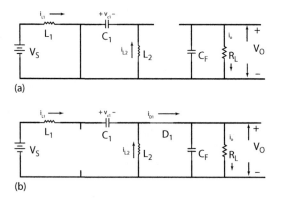

FIGURE 8.2
SEPIC converter equivalent circuits in CCM. (a) Switch ON for $0 \leq t \leq D.T$ s. (b) Switch OFF for $D.T \leq t \leq T$ s.

Interactive Models for Fourth-Order DC to DC Converters

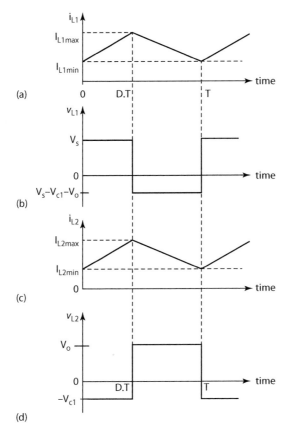

FIGURE 8.3
SEPIC converter waveforms in CCM. (a) Inductor L_1 current i_{L1}. (b) Inductor L_1 voltage vL_1. (c) Inductor L_2 current i_{L2}. (d) Inductor L_2 voltage vL_2.

Using Equations 8.4 and 8.5, we have the following:

$$I_{L1} = \frac{V_S \cdot D^2}{(1-D)^2 \cdot R_L} \tag{8.6}$$

The maximum and minimum inductor current for L_1 can be expressed as follows:

$$I_{L1max} = \frac{V_S \cdot D^2}{(1-D)^2 \cdot R_L} + \frac{V_S \cdot D \cdot T}{2L_1} \tag{8.7}$$

$$I_{L1min} = \frac{V_S \cdot D^2}{(1-D)^2 \cdot R_L} - \frac{V_S \cdot D \cdot T}{2L_1} \tag{8.8}$$

The minimum value of inductance for L_1, known as *critical inductance*, L_{1crit}, to maintain CCM is obtained by setting Equation 8.8 to zero.

$$L_{1crit} = \frac{(1-D)^2 .T.R_L}{2D} \qquad (8.9)$$

Similarly, for the inductor L_2, the maximum and minimum values of inductor current I_{L2max} and I_{L2min} can be expressed as follows:

$$I_{L2max} = \frac{V_O}{R_L} + \frac{V_O.(1-D).T}{2L_2} \qquad (8.10)$$

$$I_{L2min} = \frac{V_O}{R_L} - \frac{V_O.(1-D).T}{2L_2} \qquad (8.11)$$

The minimum value of inductance for L_2, L_{2crit}, to maintain CCM is obtained by setting Equation 8.11 to zero.

$$L_{2crit} = \frac{(1-D).T.R_L}{2} \qquad (8.12)$$

8.3 Analysis of SEPIC Converter in DCM

In DCM, the two inductor currents are not continuous. During the switch OFF period, the two inductor currents i_{L1} and i_{L2} decay and reach and maintain zero value for a brief period of time. The equivalent circuit for DCM operation is shown in Figure 8.4.

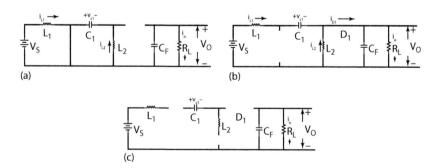

FIGURE 8.4
SEPIC converter equivalent circuits in DCM. (a) Switch ON for $0 \le t \le D.T$ s. (b) Switch OFF for $D.T \le t \le D_1.T$ s. (c) Inductor currents i_{L1} and i_{L2} are zero.

The relevant waveforms for DCM are shown in Figure 8.5 [1–5]. During the switch ON period, the two inductor currents i_{L1} and i_{L2} increase and reach maximum values I_{L1max} and I_{L2max} during the interval $0 \leq t \leq D.T$. When the switch is OFF, the two inductor currents decay and reach a value of zero during the interval $D.T \leq t \leq D_1.T$. From $D_1.T$ to T s, the two inductor currents remain zero and the cycle repeats. The diode current i_{D1} flows only during the switch OFF period from $D.T$ to $D_1.T$ s.

The average inductor current over one switching cycle is zero and, referring to Figures 8.4 and 8.5, the following equations are valid for inductor currents i_{L1} and i_{L2}:

$$\frac{V_S.D.T}{L_1} + \frac{(V_S - V_{C1} - V_O).(D_1 - D).T}{L_1} = 0 \qquad (8.13)$$

$$\frac{(-V_{C1}).D.T}{L_2} + \frac{(V_O).(D_1 - D).T}{L_2} = 0 \qquad (8.14)$$

From Equation 8.14,

$$\frac{V_{C1}}{V_O} = \frac{(D_1 - D)}{D} \qquad (8.15)$$

From Equations 8.13 and 8.15,

$$\frac{V_O}{V_S} = M = \frac{D}{(D_1 - D)} \qquad (8.16)$$

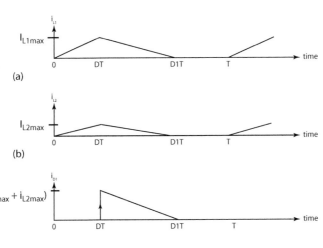

FIGURE 8.5
SEPIC converter waveforms in DCM. (a) Inductor L_1 current. (b) Inductor L_2 current. (c) Diode D_1 current.

V_O has a polarity that is positive at the top and negative at the bottom terminal, as shown in Figure 8.4.

From Figure 8.4, it is seen that the load current I_O is the average value of diode D_1 current i_{D1}, and i_{D1} flows through the load resistance only during the switch OFF period. Referring to Figure 8.5c, I_O can be expressed as follows:

$$I_O = \frac{V_O}{R_L} = \frac{1}{T} \cdot \left[\frac{(I_{L1max} - I_{L2max}).(D_1 - D).T}{2} \right] \tag{8.17}$$

$$I_O = \frac{V_O}{R_L} = \left[\frac{(D_1 - D)}{2} \right] * \left(\frac{V_S.D.T}{L_1} + \frac{V_{C1}.D.T}{L_2} \right) \tag{8.18}$$

Using Equations 8.15 and 8.16 in Equation 8.18 and simplifying gives the following result:

$$V_O = \left[\frac{V_S.D.T.R_L.(D_1 - D)}{2} \right] * \left[\frac{1}{L_1} + \frac{1}{L_2} \right] \tag{8.19}$$

$$\text{Let} \quad \frac{1}{L_{eq}} = \left[\frac{1}{L_1} + \frac{1}{L_2} \right] \tag{8.20}$$

$$\text{Let} \quad K = \frac{2.L_{eq}}{T.R_L} \tag{8.21}$$

Using Equations 8.20 and 8.21 in Equation 8.19 and simplifying gives the following:

$$\frac{V_O}{V_S} = M = \frac{D.(D_1 - D)}{K} \tag{8.22}$$

Using Equations 8.16 and 8.22, the expression for V_O/V_S can be written as given here:

$$\frac{V_O}{V_S} = M = \frac{D}{\sqrt{K}} \tag{8.23}$$

Also from Equation 8.16, D_1 can be expressed as follows:

$$D_1 = \frac{D.(1+M)}{M} \tag{8.24}$$

8.4 Model of SEPIC Converter in CCM and DCM

Models for fourth-order converters are available in references [6–8]. The switching function concept is used to model the SEPIC converter [6–8]. The SEPIC converter model parameters are shown in Table 5.1. The development of the model is given as follows:

Referring to Figures 8.2 and 8.4, the following equations can be derived for CCM and DCM operation.

$$
\begin{aligned}
i_{L1\_dot} &= \frac{(V_S)}{L_1} \quad \text{for} \quad 0 \le t \le D.T \\
&= \frac{(V_S - V_{C1} - V_O)}{L_1} \quad \text{for} \quad D.T \le t \le T
\end{aligned}
\tag{8.25}
$$

Now, for the switch S_1, the switching function SF has already been defined in Equation 5.14 for CCM operation. For the inductor current i_{L1}, another switch function SF_I_{L1} has already been defined in Equation 5.15. For the inductor current i_{L2} through L_2, the switch function SF_I_{L2} is defined as follows:

$$
\begin{aligned}
\text{SF\_}I_{L2} &= 1 \quad \text{for} \quad i_{L2} \succ 0 \\
&= 0 \quad \text{for} \quad i_{L2} \le 0
\end{aligned}
\tag{8.26}
$$

$$
\begin{aligned}
i_{L2\_dot} &= \frac{(V_{C1})}{L_2} \quad \text{for} \quad 0 \le t \le D.T \\
&= \frac{(-V_O)}{L_2} \quad \text{for} \quad D.T \le t \le T
\end{aligned}
\tag{8.27}
$$

$$
\begin{aligned}
v_{C1\_dot} &= \frac{-I_{L2}}{C_1} \quad \text{for} \quad 0 \le t \le D.T \\
&= \frac{I_{L1}}{C_1} \quad \text{for} \quad D.T \le t \le T
\end{aligned}
\tag{8.28}
$$

$$
\begin{aligned}
v_{O\_dot} &= \frac{-V_O}{(R_L.C_F)} \quad \text{for} \quad 0 \le t \le D.T \\
&= \frac{\left(I_{L1} + I_{L2} - \dfrac{V_O}{R_L}\right)}{C_F} \quad \text{for} \quad D.T \le t \le T
\end{aligned}
\tag{8.29}
$$

Load current I_O is given by Equation 5.19. In Equations 8.25, 8.27, 8.28 and 8.29, the symbol _dot indicates the first derivative with respect to time of the respective variable. These equations are valid for DCM by replacing $D.T \leq t \leq T$ by $D.T \leq t \leq D_1.T$. The diode D_1 current i_{D1} flows through the load resistor and filter capacitor during the switch OFF period in CCM and in DCM until this diode current falls to zero. This diode current i_{D1} during the switch OFF period is the algebraic sum of the two inductor currents i_{L1} and i_{L2}. The output voltage equation is formulated by observation of equivalent circuits for CCM and DCM (Figures 8.2 and 8.4), as shown in Equation 8.29.

The interactive model of the SEPIC converter suitable for CCM and DCM operation with a dialogue box is shown in Figure 8.6 and the model subsystem with a dialogue box is shown in Figure 8.7.

The data in the dialogue box in Figure 8.6 corresponds to DCM operation. For CCM operation, the relevant data are entered in the dialogue box.

The switch function SF for the switch defined in Equation 5.14 is generated using the Pulse Generator block, whose amplitude is 1, pulse width (d*100) (% of period) and switching period $1/(f_{sw})$, where d is the duty ratio and fsw is the frequency of switching. The switch function for L_1 inductor current SF_I_{L1}, defined in Equation 5.15, is generated using the embedded MATLAB® function whose input is I_{L1} and output is SF_I_{L1}, as shown in Figure 8.7.

The switch function for L_2 inductor current SF_I_{L2}, defined in Equation 8.26, is also generated in the same embedded MATLAB function, whose input is I_{L2} and output is SF_I_{L2}. Equations 8.25, 8.27, 8.28 and 8.29 are solved using another embedded MATLAB function and integrators, as shown in

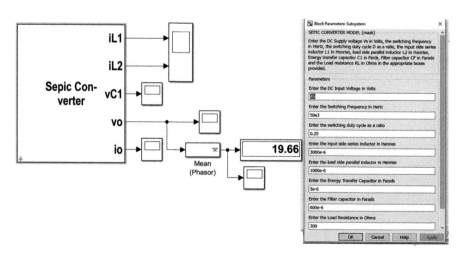

FIGURE 8.6
SEPIC converter.

Interactive Models for Fourth-Order DC to DC Converters

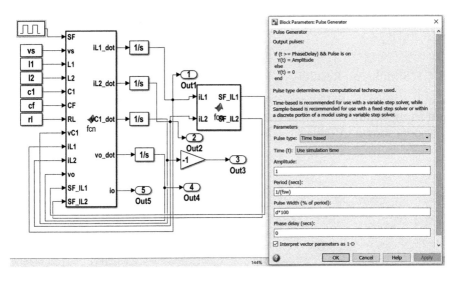

FIGURE 8.7
SEPIC converter model subsystem.

Figure 8.7. The source code for the two embedded MATLAB functions is given here:

```
function [iL1_dot,iL2_dot,vC1_dot,vo_dot,io] = fcn(SF,vs,L1,L2,
C1,CF,RL,vC1,iL1,iL2,vo,SF_IL1,SF_IL2)
%%SEPIC Converter
iL1_dot = SF*vs/(L1) + (1-SF)*SF_IL1*(-vC1+vs-vo)/(L1);
iL2_dot = SF*(vC1)/(L2) + (1-SF)*SF_IL2*(-vo)/(L2);
vC1_dot = -SF*SF_IL2*iL2/(C1) + (1-SF)*SF_IL1*iL1/(C1);
vo_dot = -SF*vo/(RL*CF) + (1-SF)*(SF_IL1*iL1+SF_IL2*iL2-
        vo/(RL))/(CF);
io = vo/(RL);
function [SF_IL1,SF_IL2] = fcn(iL1,iL2)
%%SEPIC Converter
if iL1 <= 0
        SF_IL1 = 0;
else
        SF_IL1 = 1;
end
if iL2 <= 0
        SF_IL2 = 0;
else
        SF_IL2 = 1;
end
```

The embedded MATLAB outputs v_{o_dot}, v_{C1_dot}, i_{L1_dot} and i_{L2_dot} are integrated using Integrator blocks to obtain the required outputs and are used as feedback inputs along with SF_I_{L1} and SF_I_{L2} to solve Equations 8.25, 8.27, 8.28 and 8.29.

8.4.1 Simulation Results

Simulation of the SEPIC converter for CCM and DCM operations was carried out using Simulink® [9]. The ode23t (mod. stiff/trapezoidal) solver was used. The data shown in Table 8.1 were used for simulation. The simulation results

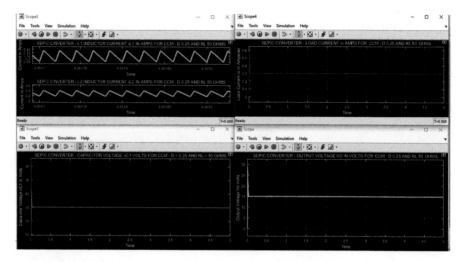

FIGURE 8.8
SEPIC converter in CCM: simulation results for $D = 0.25$.

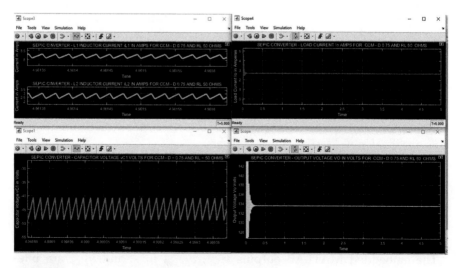

FIGURE 8.9
SEPIC converter in CCM: simulation results for $D = 0.75$.

TABLE 8.1

SEPIC Converter Model Parameters

Sl. No.	Input Voltage V_S (V)	Switching Frequency f_{sw} (Hz)	Duty Cycle D	Inductor L_1 (H)	Inductor L_2 (H)	Capacitor C_1 (F)	Filter Capacitor C_F (F)	Load Resistor R_L (Ω)	Remarks
1	45	50E3	0.25	3000E−6	1000E−6	5E−6	600E−6	50	CCM
2	45	50E3	0.75	3000E−6	1000E−6	5E−6	600E−6	50	CCM
3	45	50E3	0.25	3000E−6	1000E−6	5E−6	600E−6	300	DCM

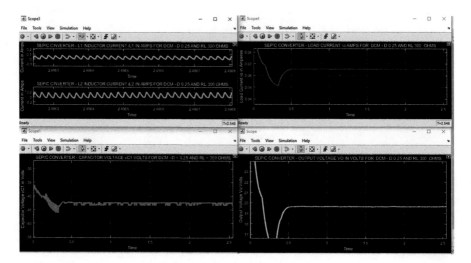

FIGURE 8.10
SEPIC converter in DCM: simulation results.

for the SEPIC converter in CCM are shown in Figures 8.8 and 8.9 and those for DCM are shown in Figure 8.10.

Simulation results and calculated values are tabulated in Tables 8.2 and 8.3.

8.5 Analysis of Quadratic Boost Converter in the CCM

The quadratic boost converter circuit schematic is shown in Figure 8.11 [10, 11].

The switch is driven by a gate pulse whose period is T and duty cycle D. The equivalent circuits of this converter in the range $0 \leq t \leq D.T$ and $D.T \leq t \leq T$ are shown in Figure 8.12 a, b, respectively. When S_1 is closed, the inductor currents find a path via D_3 to the short circuit caused by the switch, leaving D_1 and D_2 open. When S_1 is open, the inductor currents find a path through D_1 and D_2 to the load resistor. The relevant waveforms for CCM operation are shown in Figure 8.13.

Referring to Figure 8.12, noting that the average current through inductors L_1 and L_2 over one switching cycle is zero, we have the following equations:

$$\frac{V_S.D.T}{L_1} + \frac{(V_S - V_{C1}).(1-D).T}{L_1} = 0 \tag{8.30}$$

$$\frac{(V_{C1}).D.T}{L_2} + \frac{(V_{C1} - V_O).(1-D).T}{L_2} = 0 \tag{8.31}$$

TABLE 8.2

SEPIC Converter: Simulation Results

Sl. No.	Duty Ratio	Minimum L_1 Inductor Current (A)	Maximum L_1 Inductor Current (A)	Minimum L_2 Inductor Current (A)	Maximum L_2 Inductor Current (A)	Capacitor C_1 Voltage V_{C1} (V)	Output Voltage V_O (V)	Load Current I_O (A)	Remarks
1	0.25	0.0789	0.154	0.1825	0.407	−44.98	15.040	0.3007	CCM
2	0.75	8.085	8.31	2.345	3.013	−44.85	133.9	2.678	CCM
3	0.25	−0.0059	0.083	−0.0497	0.212	−42.44	19.76	0.0658	DCM

TABLE 8.3

SEPIC Converter: Calculated Values

Sl. No.	Duty Ratio	Minimum L_1 Inductor Current (A)	Maximum L_1 Inductor Current (A)	Minimum L_2 Inductor Current (A)	Maximum L_2 Inductor Current (A)	Capacitor C_1 Voltage V_{C1} (V)	Output Voltage V_O (V)	Load Current I_O (A)	Remarks
1	0.25	0.0625	0.1375	0.1875	0.4125	−45	15	0.3	CCM
2	0.75	7.9875	8.2125	2.3625	3.0375	−45	135	2.7	CCM
3	0.25	0	0.075	−0.0375	0.1875	−45	22.5	0.075	DCM

Interactive Models for Fourth-Order DC to DC Converters

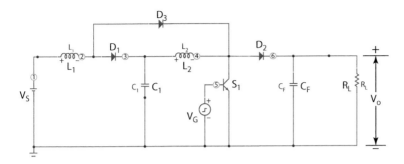

FIGURE 8.11
Quadratic boost converter.

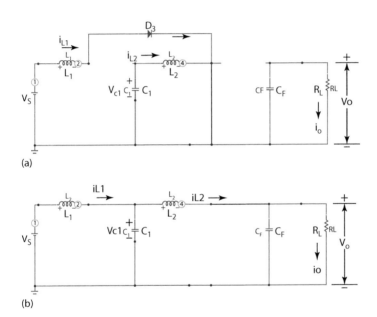

FIGURE 8.12
Quadratic boost converter equivalent in CCM. (a) Switch ON for $0 \leq t \leq D.T$ s. (b) Switch OFF for $D.T \leq t \leq T$ s.

From Equation 8.31, V_{C1} can be expressed as follows:

$$V_{C1} = V_O \cdot (1-D) \tag{8.32}$$

Using Equations 8.30 and 8.32, we have the following:

$$\frac{V_O}{V_S} = M = \frac{1}{(1-D)^2} \tag{8.33}$$

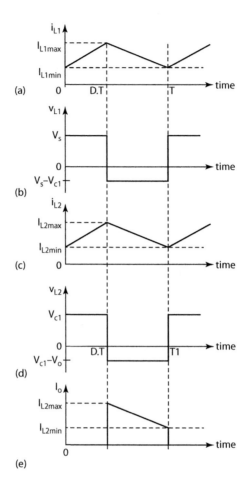

FIGURE 8.13
Quadratic boost converter waveforms in CCM. (a) Inductor L_1 current i_{L1}. (b) Inductor L_1 voltage v_{L1}. (c) Inductor L_2 current i_{L2}. (d) Inductor L_2 voltage v_{L2}. (e) Load current I_O.

Neglecting switching losses, the input power and output power are equal. This can be expressed as follows:

$$V_S * I_{L1} = V_O . I_O = \frac{V_O^2}{R_L} \tag{8.34}$$

Using Equations 8.33 and 8.34, we have the following:

$$I_{L1} = \frac{V_S}{(1-D)^4 . R_L} \tag{8.35}$$

The maximum and minimum inductor current for L_1 can be expressed as follows:

Interactive Models for Fourth-Order DC to DC Converters

$$I_{L1max} = \frac{V_S}{(1-D)^4.R_L} + \frac{V_S.D.T}{2L_1} \tag{8.36}$$

$$I_{L1min} = \frac{V_S}{(1-D)^4.R_L} - \frac{V_S.D.T}{2L_1} \tag{8.37}$$

The average voltage across C_1 is zero, which leads to the following equation:

$$\frac{I_{L2}.D.T}{C_1} + \frac{(I_{L2}-I_{L1})(1-D).T}{C_1} = 0 \tag{8.38}$$

$$I_{L2} = I_{L1}.(1-D) = \frac{V_S}{(1-D)^3.R_L} \tag{8.39}$$

$$I_{L2max} = \frac{V_S}{(1-D)^3.R_L} + \frac{V_S.D.T}{2.(1-D).L_2} \tag{8.40}$$

$$I_{L2min} = \frac{V_S}{(1-D)^3.R_L} - \frac{V_S.D.T}{2.(1-D).L_2} \tag{8.41}$$

The critical inductances L_{1crit} and L_{2crit} to maintain CCM are obtained by setting Equations 8.37 and 8.41 to zero.

$$L_{1crit} = \frac{(1-D)^4.D.T.R_L}{2} \tag{8.42}$$

$$L_{2crit} = \frac{(1-D)^2.D.T.R_L}{2} \tag{8.43}$$

8.6 Analysis of Quadratic Boost Converter in the DCM

In DCM, the two inductor currents are discontinuous. During the switch OFF period, the two inductor currents i_{L1} and i_{L2} decay and reach and maintain zero value for a brief period of time. The equivalent circuit for DCM operation is shown in Figure 8.14. The relevant waveforms for DCM operation are shown in Figure 8.15 [10, 11]. During the switch ON period, the two inductor currents i_{L1} and i_{L2} increase and reach maximum values I_{L1max} and I_{L2max} during the interval $0 <= t <= D.T$. When the switch is

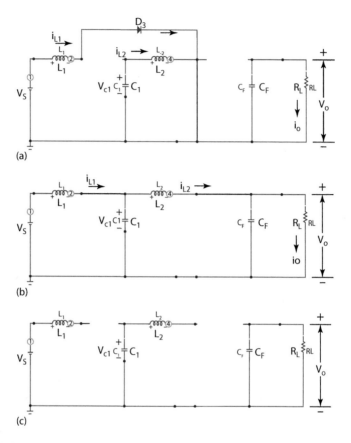

FIGURE 8.14
Quadratic boost converter in DCM. (a) Switch ON for 0 <= t <= D.T sec. (b) Switch OFF for D.T <= t <= D1.T sec. (c) Both inductor currents are zero.

OFF, the two inductor currents decay and reach a value of zero during the interval $D.T \leq t \leq D_1.T$. From $D_1.T$ to T s, the two inductor currents remain zero and the cycle repeats. The L_2 inductor current flows through the load only while the switch is OFF from $D.T$ to $D_1.T$ s. The average inductor currents i_{L1} and i_{L2} over one switching cycle are zero and, referring to Figures 8.14 and 8.15, the following equations are valid for inductor currents i_{L1} and i_{L2}:

$$\frac{V_S.D.T}{L_1} + \frac{(V_S - V_{C1}).(D_1 - D).T}{L_1} = 0 \tag{8.44}$$

$$\frac{(V_{C1}).D.T}{L_2} + \frac{(V_{C1} - V_O).(D_1 - D).T}{L_2} = 0 \tag{8.45}$$

From Equation 8.45,

$$\frac{V_{C1}}{V_O} = \frac{(D_1 - D)}{D_1} \quad (8.46)$$

From Equations 8.44 and 8.46,

$$\frac{V_O}{V_S} = M = \frac{D_1^2}{(D_1 - D)^2} \quad (8.47)$$

V_O has a polarity that is positive at the top and negative at the bottom terminal, as shown in Figure 8.11. From Figure 8.14, it is seen that the load current I_O is the average value of L_2 inductor current i_{L2} and flows through the load resistance only during the switch OFF period. Referring to Figure 8.15c, I_O can be expressed as follows:

$$I_O = \frac{V_O}{R_L} = \frac{1}{T} \cdot \left[\frac{(I_{L2\max}) \cdot (D_1 - D) \cdot T}{2} \right] \quad (8.48)$$

$$I_O = \frac{V_O}{R_L} = \left[\frac{(D_1 - D)}{2} \right] * \left(\frac{V_{C1} \cdot D \cdot T}{L_2} \right) \quad (8.49)$$

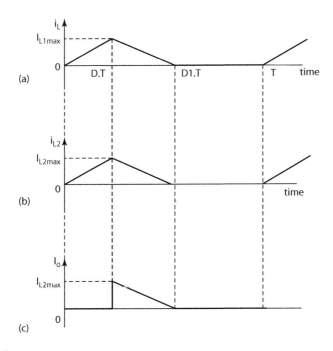

FIGURE 8.15
Quadratic boost converter waveforms in DCM. (a) Inductor L_1 current i_{L1}. (b) Inductor L_2 current i_{L2}. (c) Load current I_O.

184   *Power Electronic Converters*

Using Equations 8.46 and 8.47 in Equation 8.49 and simplifying gives the following result:

$$I_O = \frac{V_O}{R_L} = \left(\frac{V_S.D_1.D.T}{2.L_2} \right) \tag{8.50}$$

$$I_O * R_L = V_O = \left(\frac{V_S.D_1.D.T.R_L}{2.L_2} \right) \tag{8.51}$$

$$\text{Let} \quad K = \frac{2.L_2}{T.R_L} \tag{8.52}$$

$$\frac{V_O}{V_S} = M = \frac{D_1.D}{K} \tag{8.53}$$

$$D_1 = \frac{K.M}{D} \tag{8.54}$$

Using Equations 8.54 and 8.47 and simplifying,

$$M^2 * K^2 - M * \left[2 * K * D^2 + K^2 \right] + D^4 \tag{8.55}$$

Solving Equation 8.55 gives the following result:

$$M = \frac{V_O}{V_S} = \left[\frac{\left(2 * K * D^2 + K^2 \right) \pm \sqrt{\left(2 * K * D^2 + K^2 \right)^2 - 4 * K^2 * D^4}}{2 * K^2} \right] \tag{8.56}$$

Equation 8.56 gives two values for M, of which one is acceptable.

8.7 Model of Quadratic Boost Converter in CCM and DCM

The switching function concept is used to model the quadratic boost converter [6–8]. Quadratic boost converter model parameters are shown in Table 8.4. The development of the model is given here:

Referring to Figures 8.12 and 8.14, the following equations can be derived for CCM and DCM operation:

$$i_{L1\_dot} = \frac{(V_S)}{L_1} \quad \text{for} \quad 0 \le t \le D.T$$
$$= \frac{(V_S - V_{C1})}{L_1} \quad \text{for} \quad D.T \le t \le T \tag{8.57}$$

TABLE 8.4

Quadratic Boost Converter Model Parameters

Sl. No.	Input Voltage V_S (V)	Switching Frequency f_{sw} (Hz)	Duty Cycle D	Inductor L_1 (H)	Inductor L_2 (H)	Capacitor C_1 (F)	Filter Capacitor C_F (F)	Load Resistor R_L (Ω)	Remarks
1	18	50E3	0.4	90E−6	382E−6	22E−6	100E−6	50	CCM
2	18	50E3	0.4	90E−6	382E−6	22E−6	100E−6	600	DCM

Now, for the switch S_1 and inductor current i_{L1}, the switching functions SF and SF_I_{L1} have already been defined in Equations 5.14 and 5.15, respectively. For the L_2 inductor current i_{L2}, the switch function SF_I_{L2} is defined in Equation 8.26. The L_2 inductor current i_{L2}, capacitor voltage V_{C1} and output voltage V_O are defined as follows:

$$i_{L2\_dot} = \frac{(V_{C1})}{L_2} \quad \text{for} \quad 0 \leq t \leq D.T$$
$$= \frac{(V_{C1} - V_O)}{L_2} \quad \text{for} \quad D.T \leq t \leq T \quad (8.58)$$

$$v_{C1\_dot} = \frac{-I_{L2}}{C_1} \quad \text{for} \quad 0 \leq t \leq D.T$$
$$= \frac{(I_{L1} - I_{L2})}{C_1} \quad \text{for} \quad D.T \leq t \leq T \quad (8.59)$$

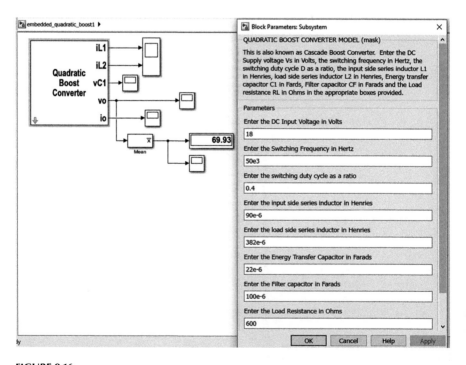

FIGURE 8.16
Quadratic boost converter.

$$v_{O\_dot} = \frac{-V_O}{(R_L . C_F)} \quad \text{for} \quad 0 \leq t \leq D.T$$

$$= \frac{\left(I_{L2} - \dfrac{V_O}{R_L}\right)}{C_F} \quad \text{for} \quad D.T \leq t \leq T \tag{8.60}$$

Load current I_O is given by Equation 5.19. In Equations 8.55 through 8.58, the symbol _dot indicates the first derivative with respect to time of the respective variable. These equations are valid for DCM by replacing $D.T <= t <= T$ by $D.T <= t <= D_1.T$. The L_2 inductor current flows through diode D_2 and the load resistor and filter capacitor during the switch OFF period in CCM and in DCM until this inductor current through L_2 falls to zero. The output voltage equation is formulated by observation of equivalent circuits for CCM and DCM (Figures 8.12 and 8.14), as shown in Equation 8.60.

The interactive model of the quadratic boost converter suitable for CCM and DCM operation with a dialogue box is shown in Figure 8.16 and the model subsystem with a dialogue box is shown in Figure 8.17.

The data in the dialogue box in Figure 8.16 corresponds to DCM operation. For CCM operation, the relevant data are entered in the dialogue box.

The switch function SF for the switch defined in Equation 5.14 is generated using the Pulse Generator block, whose amplitude is 1, pulse width (d*100) (% of period) and switching period 1/(f$_{sw}$), where d is the duty ratio and fsw

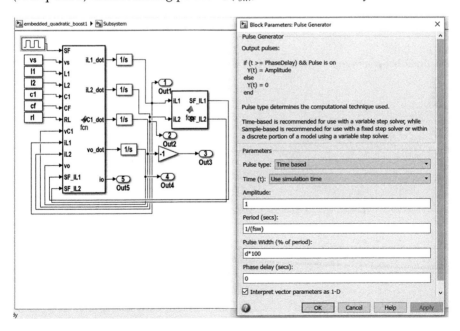

FIGURE 8.17
Quadratic boost converter model subsystem.

is the frequency of switching. The switch function for L_1 inductor current SF_I_{L1}, defined in Equation 5.15, is generated using the embedded MATLAB function, whose input is I_{L1} and output is SF_I_{L1}, as shown in Figure 8.17. The switch function for L_2 inductor current SF_I_{L2}, defined in Equation 8.26, is also generated in the same embedded MATLAB function, whose input is I_{L2} and output is SF_I_{L2}. The source code for generating SF_I_{L1} and SF_I_{L2} are the same as given in Section 8.4. Equations 8.57 through 8.60 are solved using another embedded MATLAB function and integrators as shown in Figure 8.17 and the source code is written as explained in Section 8.4.

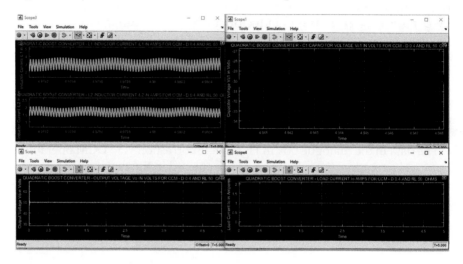

FIGURE 8.18
Quadratic boost converter in CCM for $D = 0.4$: simulation results.

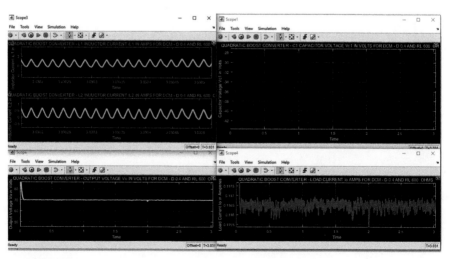

FIGURE 8.19
Quadratic boost converter in DCM for $D = 0.4$: simulation results.

TABLE 8.5

Quadratic Boost Converter: Simulation Results

Sl. No.	Duty Ratio	Minimum L_1 Inductor Current (A)	Maximum L_1 Inductor Current (A)	Minimum L_2 Inductor Current (A)	Maximum L_2 Inductor Current (A)	Capacitor C_1 Voltage V_{C1} (V)	Output Voltage V_O (V)	Load Current I_O (A)	Remarks
1	0.4	2.04	3.9	1.34	2.07	−29.9	49.9	0.9979	CCM
2	0.4	−0.204	1.58	−0.137	0.737	−35.48	70.17	0.117	DCM

TABLE 8.6

Quadratic Boost Converter: Calculated Values

Sl. No.	Duty Ratio	Minimum L_1 Inductor Current (A)	Maximum L_1 Inductor Current (A)	Minimum L_2 Inductor Current (A)	Maximum L_2 Inductor Current (A)	Capacitor C_1 Voltage V_{C1} (V)	Output Voltage V_O (V)	Load Current I_O (A)	Remarks
1	0.4	1.977	3.577	1.038	2.2944	−30	50	1.0	CCM
2	0.4	−0.064	1.536	−0.163	0.4435	−38.92	84.156	0.140	DCM

Interactive Models for Fourth-Order DC to DC Converters

The embedded MATLAB outputs v_{O_dot}, v_{C1_dot}, i_{L1_dot} and i_{L2_dot} are integrated using Integrator blocks to obtain the required outputs and are used as feedback inputs along with SF_I_{L1} and SF_I_{L2} to solve Equations 8.57 through 8.60.

8.7.1 Simulation Results

The simulation of the quadratic boost converter for CCM and DCM operations were carried out using Simulink [9]. The ode23t (mod. stiff/trapezoidal) solver was used. The data shown in Table 8.4 were used for simulation. The simulation results for the quadratic boost converter in CCM are shown in Figure 8.18 and those for DCM are shown in Figure 8.19. Simulation results and calculated values are tabulated in Tables 8.5 and 8.6.

8.8 Analysis of Ultra-Lift Luo Converter in the CCM

The ultra-lift Luo converter schematic is shown in Figure 8.20 [2, 12]. The switch S_1 is driven by a gate pulse whose period is T and duty cycle is D. The switch S_1 is ON for the interval $0 <= t <= 1D.T$ s and OFF for the interval $D.T <= t <= T$ s. The equivalent circuit of the ultra-lift Luo converter in CCM is shown in Figure 8.21. The analysis of this converter in CCM is given next:

Referring to Figure 8.21, the average inductor current through L_1 and L_2 and, similarly, the average voltage across C_1 and C_F over one switching cycle is zero. We have the following equations:

$$\frac{(V_S.D.T)}{L_1} + \frac{(V_{C1}).(1-D).T}{L_1} = 0 \qquad (8.61)$$

$$\frac{((V_S - V_{C1}).D.T)}{L_2} + \frac{(V_O - V_{C1}).(1-D).T}{L_2} = 0 \qquad (8.62)$$

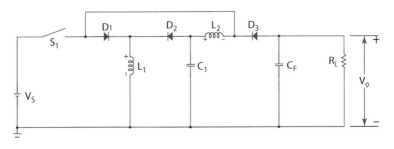

FIGURE 8.20
Ultra-lift Luo converter.

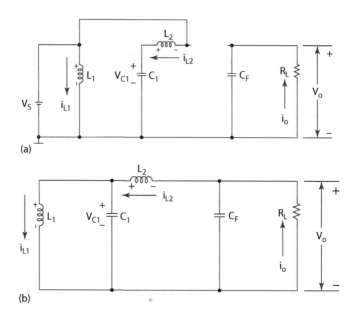

FIGURE 8.21
Ultra-lift Luo converter in CCM equivalent circuits. (a) Switch ON for $0 \leq t \leq D.T$ s. (b) Switch OFF for $D.T \leq t \leq T$ s.

From Equation 8.61,

$$\frac{V_{C1}}{V_S} = \frac{-D}{(1-D)} \tag{8.63}$$

Using Equation 8.63 in Equation 8.62 and simplifying,

$$\frac{V_O}{V_S} = \frac{-D.(2-D)}{(1-D)^2} \tag{8.64}$$

The minus sign in Equations 8.63 and 8.64 indicates that the polarities of V_{C1} and V_O in Figure 8.20 are negative at the top and positive at the bottom terminal. The waveforms of inductor currents i_{L1}, i_{L2} and load current i_O are shown in Figure 8.22 for CCM operation. The average capacitor voltage across C_1 and C_F over one switching cycle is zero. This leads to the following equations:

$$\frac{I_{L2}.D.T}{C_1} + \frac{(I_{L2}-I_{L1}).(1-D).T}{C_1} = 0 \tag{8.65}$$

$$\frac{I_O.D.T}{C_F} + \frac{(I_O-I_{L2}).(1-D).T}{C_F} = 0 \tag{8.66}$$

Interactive Models for Fourth-Order DC to DC Converters

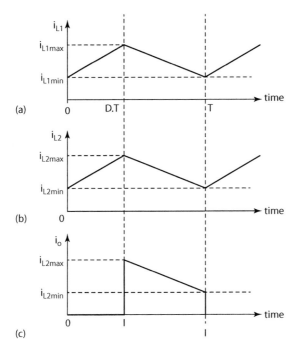

FIGURE 8.22
Ultra-lift Luo converter waveforms in CCM. (a) Inductor L_1 current i_{L1}. (b) Inductor L_2 current i_{L2}. (c) Load current I_O.

From Equations 8.65 and 8.66,

$$\frac{I_{L1}}{I_{L2}} = \frac{1}{(1-D)} \tag{8.67}$$

$$I_{L2} = \frac{I_O}{(1-D)} = \frac{V_O}{R_L \cdot (1-D)} \tag{8.68}$$

Using Equation 8.68 in Equation 8.67,

$$I_{L1} = \frac{I_O}{(1-D)^2} = \frac{V_O}{R_L \cdot (1-D)^2} \tag{8.69}$$

Using Equation 8.64 in Equations 8.68 and 8.69,

$$I_{L2} = \frac{V_S \cdot D \cdot (2-D)}{R_L \cdot (1-D)^3} \tag{8.70}$$

$$I_{L1} = \frac{V_S \cdot D \cdot (2-D)}{R_L \cdot (1-D)^4} \tag{8.71}$$

The maximum and minimum value of current through inductor L_1 and L_2 can be expressed as follows:

$$I_{L1\max} = \frac{V_S.D.(2-D)}{R_L.(1-D)^4} + \left|\frac{V_S.D.T}{2.L_1}\right.$$

$$I_{L1\min} = \frac{V_S.D.(2-D)}{R_L.(1-D)^4} - \left.\frac{V_S.D.T}{2.L_1}\right| \tag{8.72}$$

$$I_{L2\max} = \frac{V_S.D.(2-D)}{R_L.(1-D)^3} + \left|\frac{V_S.D.T}{2.L_2.(1-D)}\right.$$

$$I_{L2\min} = \frac{V_S.D.(2-D)}{R_L.(1-D)^3} - \left.\frac{V_S.D.T}{2.L_2.(1-D)}\right| \tag{8.73}$$

Equating $I_{L1\min}$ and $I_{L2\min}$ in Equations 8.72 and 8.73 to zero, the critical inductances L_{1_crit} and L_{2_crit} can be expressed as follows:

$$L_{1\_crit} = \frac{V_S.D.T.R_L.(1-D)^4}{2.D.(2-D)} \tag{8.74}$$

$$L_{2\_crit} = \frac{V_S.D.T.R_L.(1-D)^2}{2.D.(2-D)} \tag{8.75}$$

Equations 8.74 and 8.75 give the minimum value of inductance for L_1 and L_2 to maintain CCM.

8.9 Analysis of Ultra-Lift Luo Converter in DCM

In DCM, the inductor currents i_{L1} and i_{L2} decay, reach zero during the switch OFF period and maintain a zero value until the switch is ON for the next cycle. The equivalent circuits for DCM operation are shown in Figure 8.23. The relevant waveforms for DCM operation are shown in Figure 8.24 [2, 12].

Referring to Figure 8.23, the average inductor currents i_{L1} and i_{L2} over one switching cycle is zero, which can be expressed as follows:

$$\frac{V_S.D.T}{L_1} + \frac{V_{C1}.(D_1-D).T}{L_1} = 0 \tag{8.76}$$

$$\frac{(V_S-V_{C1}).D.T}{L_2} + \frac{(V_0-V_{C1}).(D_1-D).T}{L_2} = 0 \tag{8.77}$$

Interactive Models for Fourth-Order DC to DC Converters 195

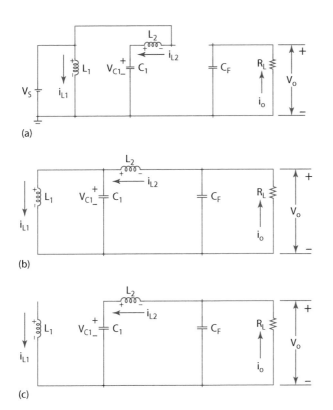

FIGURE 8.23
Ultra-lift Luo converter in DCM: equivalent circuits. (a) Switch ON for $0 \leq t \leq D.T$ s. (b) Switch OFF for $D.T \leq t \leq D_1.T$ s. (c) DCM operation.

From Equation 8.76, V_{C1} can be expressed as follows:

$$V_{C1} = \frac{-V_S.D}{(D_1 - D)} \tag{8.78}$$

Using Equation 8.78 in 8.77 and simplifying,

$$\frac{V_O}{V_S} = \frac{-D.(2D_1 - D)}{(D_1 - D)^2} \tag{8.79}$$

Referring to Figures 8.23a and b, the inductor current i_{L2} that flows during switch OFF contributes to the load voltage, V_O. Referring to Figure 8.24b, the maximum inductor current $i_{L2\max}$ can be expressed as follows:

$$i_{L2\max} = \frac{(V_S - V_{C1}).D.T}{L_2} \tag{8.80}$$

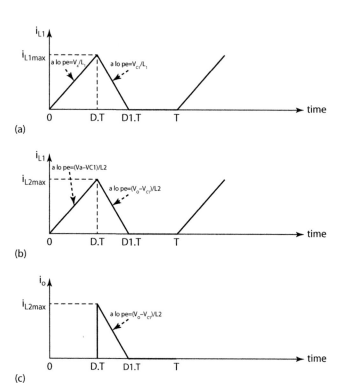

FIGURE 8.24
Ultra-lift Luo converter in DCM. (a) Inductor L_1 current iL_1. (b) Inductor L_2 current iL_2. and (c) Load current I_O.

Using Equation 8.78 in Equation 8.80,

$$i_{L2\max} = \frac{V_S.D_1.D.T}{(D_1-D).L_2} \tag{8.81}$$

Referring to Figure 8.24c, the average load current I_O can be expressed as follows:

$$I_O = \left[\frac{1}{T} * \frac{1}{2}.(D_1-D).T.\frac{V_S.D.D_1.T}{(D_1-D).L_2}\right]$$

$$= \frac{V_S.D.D_1.T}{2.L_2} \tag{8.82}$$

$$V_O = -I_O * R_L = \frac{-V_S.D.D_1.T.R_L}{2.L_2} \tag{8.83}$$

Let $\quad K = \dfrac{2.L_2}{T.R_L} \tag{8.84}$

Interactive Models for Fourth-Order DC to DC Converters 197

Using Equation 8.84 in Equation 8.83,

$$\frac{V_O}{V_S} = M = \frac{D.D_1}{K} \tag{8.85}$$

$$D_1 = \frac{K.M}{D} \tag{8.86}$$

Using Equation 8.86 in Equation 8.79,

$$\frac{V_O}{V_S} = M = \frac{2.K.M.D^2 - D^4}{\left[K^2.M^2 + D^4 - 2.M.K.D^2\right]} \tag{8.87}$$

Simplifying Equation 8.87,

$$M^3.K^2 - 2.M^2.K.D^2 + M.[D^4 - 2.K.D^2] + D^4 = 0 \tag{8.88}$$

The solution of Equation 8.88 gives three values for M, of which one is acceptable.

To calculate D_1, first find the acceptable value M_1 for M from Equation 8.88. Then use this M_1 in Equation 8.79 to find acceptable value for D_1 whose range is $D < D_1 \leq 1$.

8.10 Model of Ultra-Lift Luo Converter in CCM and DCM

The model of the ultra-lift Luo converter is reported in references [2, 12]. Here, the switching function concept is used to develop the model [6–8]. The data used for the model is shown in Table 8.7. The development of the model is given as follows:

TABLE 8.7

Ultra-Lift Luo Converter Model Parameters

Parameter	Value
Supply voltage V_s	10 V
Switching frequency f_{sw}	50 kHz
Duty cycle D used	0.6
Inductor L_1	1 mH
Inductor L_2	1 mH
E.T. capacitor C_1	1 µF
Filter capacitor C_F	1 µF
Load resistance R_L	300 Ω for CCM
	3000 Ω for DCM

Referring to Figures 8.21 and 8.23, the following equations can be derived for CCM and DCM operation.

$$i_{L1\_dot} = \frac{(V_S)}{L_1} \quad \text{for} \quad 0 \le t \le D.T$$

$$\frac{(V_{C1})}{L_1} \quad \text{for} \quad D.T \le t \le T \qquad (8.89)$$

Now, for the switch S_1 and inductor current i_{L_1}, the switching functions SF and SF_I_{L1} have already been defined in Equations 5.14 and 5.15, respectively. For the L_2 inductor current i_{L2}, the switch function SF_I_{L2} has been defined in Equation 8.26. The L_2 inductor current i_{L2}, capacitor voltage V_{C1} and output voltage V_O are defined as follows:

$$i_{L2\_dot} = \frac{(V_S - V_{C1})}{L_2} \quad \text{for} \quad 0 \le t \le D.T$$

$$= \frac{(V_O - V_{C1})}{L_2} \quad \text{for} \quad D.T \le t \le T \qquad (8.90)$$

$$v_{C1\_dot} = \frac{I_{L2}}{C_1} \quad \text{for} \quad 0 \le t \le D.T$$

$$= \frac{(I_{L2} - I_{L1})}{C_1} \quad \text{for} \quad D.T \le t \le T \qquad (8.91)$$

$$v_{O\_dot} = \frac{-V_O}{(R_L.C_F)} \quad \text{for} \quad 0 \le t \le D.T$$

$$= \frac{\left(I_{L2} - \dfrac{V_O}{R_L}\right)}{C_F} \quad \text{for} \quad D.T \le t \le T \qquad (8.92)$$

Load current I_O is given by Equation 5.19. In Equations 8.89 through 8.92, the symbol _dot indicates the first derivative with respect to time of the respective variable. These equations are valid for DCM by replacing $D.T \le t \le T$ by $D.T \le t \le D_1.T$. The L_2 inductor current flows through the load resistor and filter capacitor during the switch OFF period in CCM and in DCM until this inductor current through L_2 falls to zero. The output voltage equation is formulated by observation of equivalent circuits for CCM and DCM (Figures 8.21 and 8.23), as shown in Equation 8.92.

Interactive Models for Fourth-Order DC to DC Converters 199

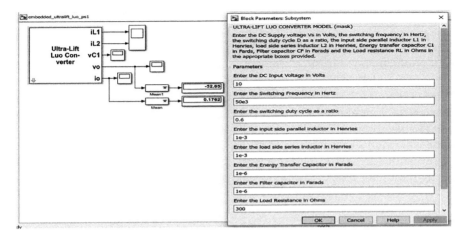

FIGURE 8.25
Ultra-lift Luo converter model.

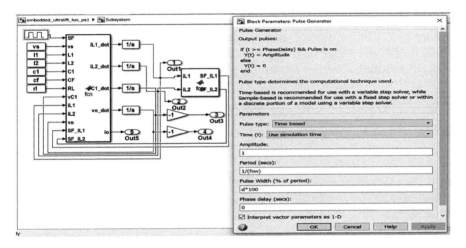

FIGURE 8.26
Ultra-lift Luo converter model subsystem.

The interactive model of the ultra-lift Luo converter suitable for CCM and DCM operation with a dialogue box is shown in Figure 8.25 and the model subsystem with a dialogue box is shown in Figure 8.26.

The data in the dialogue box in Figure 8.25 corresponds to CCM operation. For DCM operation, the relevant data, 3000 Ω for load resistance, is entered in the dialogue box.

The switch function SF for the switch defined in Equation 5.14 is generated using the Pulse Generator block, whose amplitude is 1, pulse width (d*100) (% of period) and switching period 1/(fsw) where d is the duty ratio and fsw is the frequency of switching.

The switch function for the L_1 inductor current SF_I_{L1} defined in Equation 5.15 is generated using the embedded MATLAB function whose input is I_{L1} and output is SF_I_{L1}, as shown in Figure 8.26. The switch function for L_2 inductor current SF_I_{L2} defined in Equation 8.26 is also generated in the same embedded MATLAB function whose input is I_{L2} and output is SF_I_{L2}. The source code for generating SF_I_{L1} and SF_I_{L2} are the same as given in Section 8.4. Equations 8.89 through 8.92 are solved using another embedded MATLAB function and integrators as shown in Figure 8.26 and the source code is written as explained in Section 8.4.

The embedded MATLAB outputs v_{O_dot}, v_{C1_dot}, i_{L1_dot} and i_{L2_dot} are integrated using Integrator blocks to obtain the required outputs and are used as feedback inputs along with SF_I_{L1} and SF_I_{L2} to solve Equations 8.89 through 8.92.

8.10.1 Simulation Results

The simulation of the ultra-lift Luo converter is carried out using the ode23t (mod. stiff/trapezoidal) solver in Simulink [9]. The MATLAB program to solve Equation 8.88 is shown in Program Segment 8.1 and the result is in Figure 8.27. The simulation results for CCM are shown in Figure 8.28 and those for DCM are shown in Figure 8.29. The simulation results and the theoretically calculated values are tabulated in Tables 8.8 and 8.9 respectively.

```
%%Program Segment 8.1
%%Ultra-Lift Luo Converter in DCM, %%Program author: Dr.
Narayanaswamy.P.R.Iyer
vs = 10;
l1 = 1e-3;
l2 = 1e-3;
c1 = 1e-6;
cf = 1e-6;
rl = 3000;
tsw = 1/(50e3);
d = 0.6;
k = (2*l2)/(tsw*rl);
%%function f(m) = M^(3)*K^(2) - 2.M^(2)*D^(2)*K + M*(D^(4)
                - 2*K*D^(2)) + D^(4)
%%Let M be the voltage gain Vo/Vs
for M = 1:15,
    m = M;
    x(m) = m;
    f(m) = (m^(3))*(k^(2)) - 2*(m^(2))*(d^(2))*k
            + m*((d^(4)) - 2*k*(d^(2))) + d^(4);
end
plot(x,f,'-k'),grid;
xlabel('Voltage gain M');
ylabel('Function f');
title('Ultra-lift Luo Converter in DCM');
```

From Figure 8.27, the acceptable value of voltage gain M, which makes function f zero, is 8.2. The value of D_1 calculated using Equation 8.86 is not acceptable. The acceptable value M_1 of M is 8.2, obtained by solving Equation 8.88. This value of M_1 is substituted in Equation 8.79 to find the acceptable value of D_1, which is found to be 0.8951.

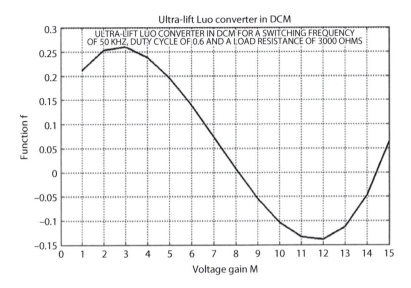

FIGURE 8.27
Ultra-lift Luo converter: simulation of Program Segment 8.1.

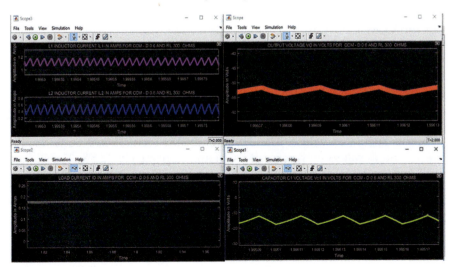

FIGURE 8.28
Ultra-lift Luo converter in CCM: simulation results.

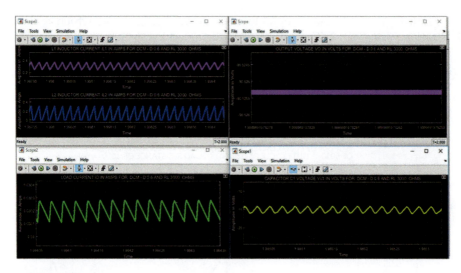

FIGURE 8.29
Ultra-lift Luo converter in DCM: simulation results.

8.11 Conclusions

Interactive system models for the SEPIC, quadratic boost and ultra-lift Luo converters are presented in this chapter. All the models developed in this chapter can be used for both CCM and DCM.

Simulation results and calculated values for the SEPIC converter in Tables 8.2 and 8.3 show that the deviation of simulation results for I_{L1min}, I_{L1max}, I_{L2min}, I_{L2max}, V_{C1}, V_O and I_O from the calculated values using formulae are very small for both CCM and DCM.

In the case of the quadratic boost converter, the simulation values such as I_{L1min} and I_{L1max} for CCM almost agree with the calculated values. Similarly, I_{L2min} and I_{L2max} for CCM have a small discrepancy with the calculated values. The simulation and calculated values for capacitor voltage V_{C1}, output voltage V_O and the load current for the case of CCM agree well. For DCM, shown in Tables 8.5 and 8.6, these values by simulation differ by a small percentage from the calculated values. The deviation of the simulation results for the V_{C1}, V_O and I_O values in the case of DCM are, respectively, 12.5%, 16.67% and 14.3% from calculated results.

For the ultra-lift Luo converter, the simulation results for i_{L1min}, i_{L1max}, i_{L2min}, i_{L2max}, V_{C1}, V_O and I_O agree closely with the calculated values in the case of CCM, and all these simulation results except V_{C1} agree closely with the calculated values in the case of DCM shown in Tables 8.8 and 8.9. In DCM, the value of V_{C1} by simulation differs by around 25% from the calculated value. Similarly, the simulation result for V_O for DCM differs by 10.9% from the calculated value.

TABLE 8.8

Ultra-lift Luo Converter: Simulation Results

Sl. No.	Duty Ratio	Minimum L_1 Inductor Current (A)	Maximum L_1 Inductor Current (A)	Minimum L_2 Inductor Current (A)	Maximum L_2 Inductor Current (A)	Capacitor C_1 Voltage V_{C1} (V)	Output Voltage V_O (V)	Load Current I_O (A)	Remarks
1	0.6	1.055	1.175	0.293	0.597	−15.29	−52.9	0.1763	CCM
2	0.6	0.241	0.366	−0.0055	0.306	−15.43	−90.7	0.032	DCM

TABLE 8.9

Ultra-lift Luo Converter: Calculated Values

Sl. No.	Duty Ratio	Minimum L_1 Inductor Current (A)	Maximum L_1 Inductor Current (A)	Minimum L_2 Inductor Current (A)	Maximum L_2 Inductor Current (A)	Capacitor C_1 Voltage V_{C1} (V)	Output Voltage V_O (V)	Load Current I_O (A)	Remarks
1	0.6	1.034	1.154	0.287	0.587	−15	−52.5	0.175	CCM
2	0.6	0.254	0.374	−0.1107	0.296	−20.33	−82.00	0.0273	DCM

References

1. B. Choi: "Pulse width modulated DC to DC power conversion: Circuit, dynamics and control designs", Piscataway, NJ: IEEE Press – Wiley, 2013; pp. 93–143.
2. F.L. Luo and H. Ye: *Power Electronics*, Boca Raton, FL: CRC Press, 2013.
3. D.S.L. Simonetti, J. Sebastian, F.S. Dos Reis, and J. Uceda: "Design criteria for SEPIC and Cuk converters as power factor preregulators in discontinuous conduction mode", *Proceedings of PEMC*, USA, 1992; pp. 283–288.
4. D.S.L. Simonetti, J. Sebastian, and J. Uceda: "The discontinuous conduction mode SEPIC and Cuk power factor preregulators: Analysis and design", *IEEE Transactions on Industrial Electronics*; Vol.44, No.5, October 1997; pp. 630–637.
5. L.G. de Vicufia, F. Guinjoan, J. Majo, and L. Martinez: "Discontinuous conduction mode in the SEPIC converter", *MELECON*, Lisbon, Portugal, 1989; pp. 38–42.
6. S. Cuk and R.D. Middlebrook: "A general unified approach to modelling switching DC-to-DC converters in discontinuous conduction mode", *Power Electronics Specialists Conference*, California, 1977; pp. 36–57.
7. H.Y. Kanaan and K. Al-Haddad: "Modeling and simulation of DC – DC power converters in CCM and DCM using the switching functions approach: Applications to the buck and Cuk converters", *IEEE-PEDS*, Malaysia, November–December 2005; pp. 468–473.
8. H.Y. Kanaan, K. Al-Haddad, and F. Fnaiech: "Switching-function-based modeling and control of a SEPIC power factor correction circuit operating in continuous and discontinuous current modes", *IEEE International Conference on Industrial Technology*, Tunisia, 2004; pp. 431–437.
9. The Mathworks Inc.: "MATLAB/Simulink Release Notes, R2016b", 2016.
10. L.H.S.C. Barreto et al.: "An optimal lossless commutation quadratic PWM boost converter", *IEEE Conference Record*, APEC 2002, Dallas, TX; pp. 624–629.
11. F.F. Tofoli et al.: "Association of a quadratic boost converter and a new topology of soft-switched two switch forward converter", *35th IEEE Power Electronics Specialists Conference*: Aachen, Germany, 2004; pp. 633–637.
12. F.L. Luo and H. Ye: "Ultra-lift Luo converter", *IEE Proceedings on Electric Power Applications*; Vol. 152, 2005; pp. 27–32.

9

Interactive Models for Three-Phase Multilevel Inverters

9.1 Introduction

Multilevel inverters were proposed to obtain higher output voltages without the use of step-up transformers closely following a sinusoidal waveform. Additionally, the harmonic content in the output voltage waveform is greatly reduced in comparison with conventional two-level inverters [1–6]. Hence, they find wide applications in STATCOMs, adjustable speed drives (ASDs) and so on. This chapter describes the interactive system modelling of the three-phase diode-clamped three-level inverter (DCTLI) and the three-phase flying-capacitor three-level inverter (FCTLI). The highest number of levels obtainable with diode-clamped and flying-capacitor multilevel inverters are limited by factors such as the number and voltage ratings of the clamping diodes and capacitors, respectively. In general, if m is the number of voltage levels in the phase-to-ground voltage of the three-phase multilevel inverter, then the number of voltage levels in the line-to-line voltage will be $(2.m-1)$ and the number of switching pulses will be $6*(m-1)$ [2–4]. Thus, for a three-phase DCTLI or FCTLI, the number of voltage levels for the line-to-line voltage will be five and the number of gate pulses required (i.e. semiconductor switches) will be twelve.

Another multilevel inverter topology is the cascaded multilevel inverter (CHB). In this topology, several of the conventional single-phase four-switch H-bridge inverters are connected in cascade to obtain higher output voltage, each H-bridge inverter with a separate DC source. The DC source for each H-bridge inverter can be equal or unequal. In the former case, it is called a *symmetrical cascade H-bridge (SCHB) inverter*, and in the latter, it is called an *unsymmetrical cascaded H-bridge (UCHB) inverter*. The total output voltage is the algebraic sum of the output voltage of each H-bridge inverter cell. The number of output phase-to-ground voltage levels m in a CHB inverter is defined by $m=(2*s+1)$ where s is the number of separate

207

DC sources or the H-bridge inverter cell. Also, CHB inverter topology can be combined with DCTLI and FCTLI topology to form a new multilevel inverter topology.

As for a conventional inverter, several pulse width modulation (PWM) schemes are available for multilevel inverters such as the multi-carrier sine PWM (MSPWM), selective harmonic elimination PWM (SHE-PWM) and space vector PWM (SVPWM). Also, the MSPWM is classified into two groups, namely, multi-carrier level-shift PWM (MLSPWM) and multi-carrier phase-shift PWM (MPSPWM).

In this chapter, interactive system models are developed for three-phase DCTLI, FCTLI, the SCHB inverter, MLSPWM and MPSPWM.

9.2 Three-Phase Diode-Clamped Three-Level Inverter

The circuit schematic of the three-phase DCTLI is shown in Figure 9.1 using Microcap11 software [1–6,7]. There are four switches per phase, two at the top leg and two at the bottom leg. Two diodes in series are connected in between the top and bottom switches. The two capacitors at the input act as a potential divider. The DC link voltage is $2.V_{dc}$ V. The simulation results for the three-phase line-to-ground and line-to-line voltages are shown in Figure 9.2. As can be seen from Figure 9.2, the line-to-ground voltage has three voltage levels and the line-to-line voltage has five voltage levels.

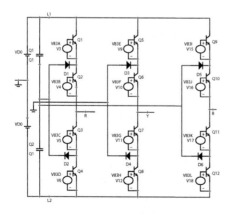

FIGURE 9.1
Three-phase DCTLI.

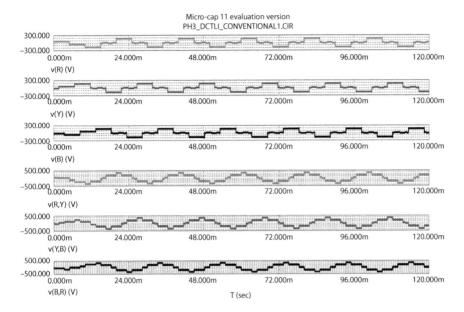

FIGURE 9.2
Three-phase DCTLI: output voltage waveforms.

9.2.1 Modelling of Three-Phase Diode-Clamped Three-Level Inverter

The model of the three-phase DCTLI is shown in Figure 9.3 [8]. The dialogue boxes are shown in Figure 9.4. The various subsystems are shown in Figure 9.5a,b. A brief description of the various subsystems are given here:

Figure 9.5a corresponds to the dialogue box 'Three-Phase DCTLI Gate Drive' shown in Figure 9.4. The switching frequency of the inverter in hertz is entered in the box provided. The output of gate drives marked Out2, Out6, Out9 and Out12 in Figure 9.5a correspond to the switches Q1, Q2, Q3 and Q4 in Figure 9.1. Similarly, the outputs of gate drives marked Out1, Out5, Out8 and Out10 correspond to switches Q5, Q6, Q7 and Q8, and the gate drive outputs marked Out3, Out4, Out7 and Out11 correspond to switches Q9, Q10, Q11 and Q12 in Figure 9.1. The switching table for the switches Q1, Q2, Q3 and Q4 in Phase R is shown in Table 9.1. The same table can be used for switches in Phases Y and B by replacing Q1 to Q4 by Q5 to Q8 or Q9 to Q12 in respective order. The start point of the appropriate phases may be noted from the remarks column of Table 9.1. If d is the time delay between any two neighbouring or consecutive states in Table 9.1, then this delay d is $\pi/6$ rad. Table 9.1 is implemented as shown in Figure 9.5a using the Pulse Generator block. The amplitude of all the pulses in the Pulse Generator block is set to 1 and the period is set to [1/fsw], where fsw is the switching frequency of the inverter. The phase delay for each of the switches in Phases R, Y and B are tabulated in Table 9.2. The switch names are in Figure 9.1. In Table 9.1,

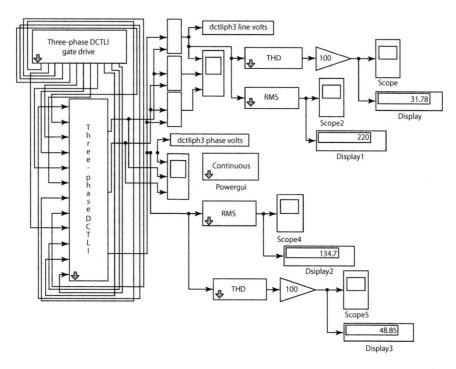

FIGURE 9.3
Three-phase DCTLI model.

gate pulse Q1 is HIGH (logic 1) for three entries and LOW (logic 0) for nine entries. Hence, its duty cycle is 3/12 (($3*\pi/6$)/($12*\pi/6$)) or 0.25. The duty cycle for the remaining switches can be calculated similarly from Table 9.1.

Figure 9.5b corresponds to the dialogue box 'Three-Phase DCTLI Unit' in Figure 9.4. Here, the DC link voltage is entered in volts in the box provided. The model of the semiconductor switches shown in Figure 9.1 is implemented using threshold switches and summing blocks as shown in Figure 9.5b. The switches Q1, Q2, Q3 and Q4 in Phase R are implemented as per Table 9.1, according to the following modelling statement:

```
If (gate pulse for Q1 is HIGH AND Q2 is HIGH) then
            Output V_rg = +V_dc/2
Else if (gate pulse for Q2 is HIGH AND Q3 is HIGH) then
            Output V_rg = 0
Else if (gate pulse for Q3 is HIGH AND Q4 is HIGH) then
            Output V_rg = -V_dc/2
            End if
```

Similar modelling statements hold good for switches in Phases Y and B. The AND gate is implemented using a threshold switch, as shown in

Interactive Models for Three-Phase Multilevel Inverters 211

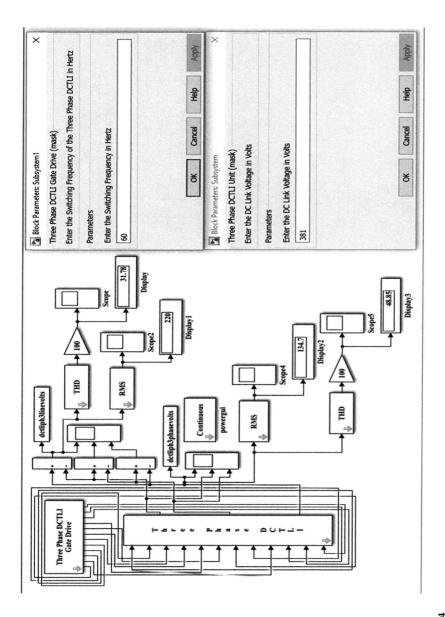

FIGURE 9.4
Three-phase DCTLI model with interactive blocks.

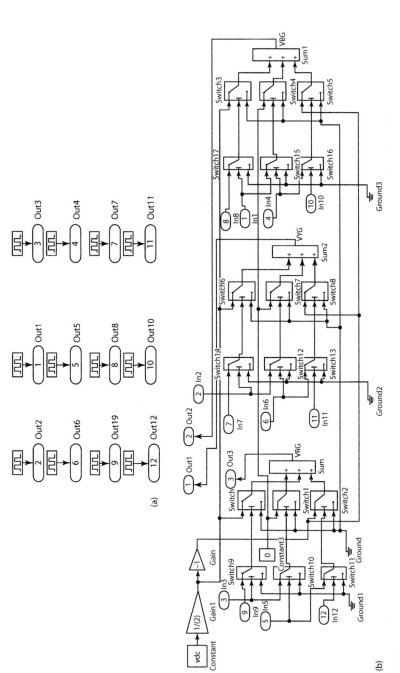

FIGURE 9.5.
(a–c) Three-phase DCTLI model subsystems.

Interactive Models for Three-Phase Multilevel Inverters

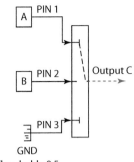

Threshold - 0.5
Criteria for passing first input u(2) >= 0.5
(c)

FIGURE 9.5. (CONTINUED)

TABLE 9.1

Switching Table for Three-Phase DCTLI

Q1	Q2	Q3	Q4	V_{rg}	Remarks
1	1	0	0	$+V_{dc}/2$	Phase R start
1	1	0	0	$+V_{dc}/2$	
1	1	0	0	$+V_{dc}/2$	
0	1	1	0	0	
0	1	1	0	0	Phase B start
0	1	1	0	0	
0	0	1	1	$-V_{dc}/2$	
0	0	1	1	$-V_{dc}/2$	
0	0	1	1	$-V_{dc}/2$	Phase Y start
0	1	1	0	0	
0	1	1	0	0	
0	1	1	0	0	
1	1	0	0	$+V_{dc}/2$	Phase R start for next cycle

TABLE 9.2

Three-Phase DCTLI: Gate Pulse Phase Delay

Sl. No.	Switch Phase R	Phase Delay (s)	Switch Phase Y	Phase Delay (s)	Switch Phase B	Phase Delay (s)
1	Q1	0	Q5	4/(12.fsw)	Q9	8/(12.fsw)
2	Q2	9/(12.fsw)	Q6	1/(12.fsw)	Q10	5/(12.fsw)
3	Q3	3/(12.fsw)	Q7	7/(12.fsw)	Q11	11/(12.fsw)
4	Q4	6/(12.fsw)	Q8	10/(12.fsw)	Q12	2/(12.fsw)

Figure 9.5c. In Figure 9.5c, the $u(1)$ and $u(2)$ are the inputs A and B with a value of either logic 0 or 1. The $u(3)$ input is the ground or logic 0. The threshold value of the switch is 0.5. When $u(2)$ is greater than or equal to the threshold value, the output corresponds to the $u(1)$ input, else the output is the $u(3)$ input. Thus, Figure 9.5c implements the AND gate. This AND gate principle is used in Figure 9.5b. Referring to the threshold switches from top to bottom on the extreme left of Figure 9.5b, the top switch marked Switch9 performs the AND operation of the gate pulse for Q1 and Q2, the middle switch marked Switch10 performs the AND operation of the gate pulse for Q2 and Q3 and the bottom switch marked Switch11 performs the AND operation of the gate pulse for Q3 and Q4. The top, middle and bottom switch outputs on the extreme left are passed to the $u(2)$ inputs of the second column of the top, middle and bottom switches marked Switch, Switch1 and Switch2, respectively. The $u(1)$ inputs of Switch, Switch1 and Switch2 in the second column are $+V_{dc}/2$, zero and $-V_{dc}/2$, respectively, while their $u(3)$ inputs are all zeros. All these switches have a threshold value of 0.5 and output $u(1)$ when the $u(2)$ input is greater than or equal to the threshold value, else the output is the $u(3)$ input and performs the modelling statement given. The output of the second column of switches marked Switch, Switch1 and Switch2 are added in a summing block to phase-to-ground output voltage V_{rg}. The switches for Phases Y and B are developed in the same way.

9.2.2 Simulation Results

The simulation results of the line-to-line voltage, line-to-ground voltage, root mean square (RMS) value of the line-to-line and line-to-ground voltages and their total harmonic distortion (THD) are shown in Figure 9.6 for a three-phase 60 Hz DCTLI using the ode23t (mod. stiff/trapezoidal) solver [9]. The DC link voltage is 381 V. The harmonic spectrum of the line-to-line voltages is shown in Figure 9.7a,b, respectively. Simulation results are tabulated in Table 9.3.

9.3 Three-Phase Flying-Capacitor Three-Level Inverter

The circuit schematic of the three-phase 50 Hz FCTLI is shown in Figure 9.8 using Microcap11 software [1–6,7]. There are four switches per phase, two at the top leg and two at the bottom leg. The capacitor connecting the switches in each of the three phases is called the *flying capacitor*. The two capacitors at the input act as a potential divider. The DC link voltage is $2V_{dc}$ V. The simulation results for the three-phase line-to-ground and line-to-line voltages are shown in Figure 9.9. As can be seen from Figure 9.9, the line-to-ground voltage has three voltage levels and the line-to-line voltage has five voltage levels, and there are 12 gate pulses, that is, semiconductor switches.

Interactive Models for Three-Phase Multilevel Inverters 215

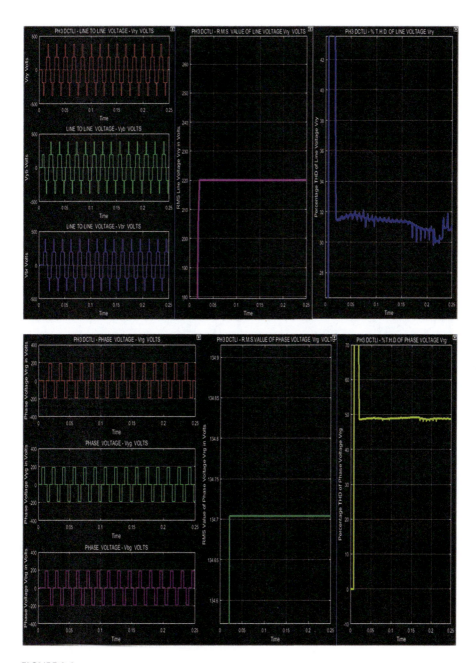

FIGURE 9.6
Three-phase DCTLI simulation results.

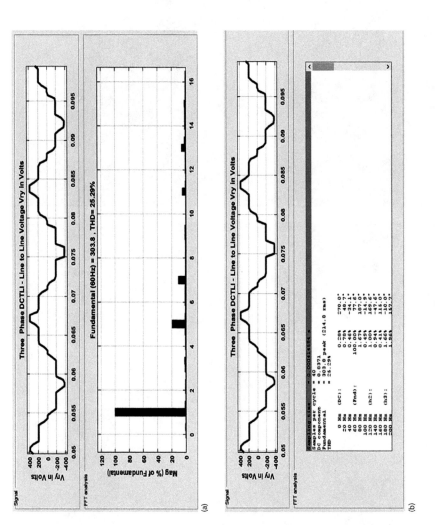

FIGURE 9.7
(a and b) Harmonic spectrum of line-to-line voltage.

Interactive Models for Three-Phase Multilevel Inverters 217

TABLE 9.3

Three-Phase DCTLI: Simulation Results

Sl. No.	DC Link Voltage (V)	Frequency (Hz)	V_{LL} RMS (V)	V_{LL} THD (%)	V_{LG} RMS (V)	V_{LG} THD (%)
1	381	60	220	30.82	134.7	48.84

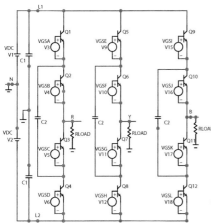

FIGURE 9.8
Three-phase FCTLI.

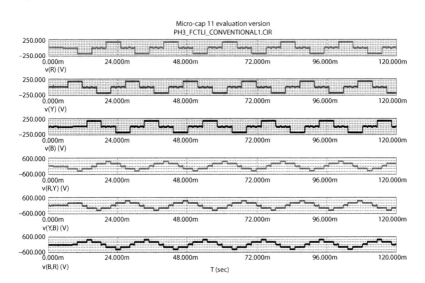

FIGURE 9.9
Three-phase FCTLI: output voltage waveforms.

9.3.1 Modelling of Three-Phase Flying-Capacitor Three-Level Inverter

The model of the three-phase FCTLI is shown in Figure 9.10 [8]. The FCTLI model, with an interactive dialogue box, is shown in Figure 9.11. The subsystem is shown in Figure 9.12. A brief description of the subsystem is given as follows:

Figure 9.11 corresponds to the dialogue box 'Three-Phase FCTLI and Gate Drive Unit' shown in Figure 9.12. The DC link voltage in volts and the switching frequency of the inverter in hertz are entered in the boxes provided. The outputs of the gate drives marked from Q1 to Q12 in Figure 9.12 correspond to the switches Q1 to Q12 in Figure 9.8. The switching table for the switches Q1, Q2, Q3 and Q4 in Phase R is shown in Table 9.4. The same table can be used for switches in Phases Y and B by replacing Q1 to Q4 by

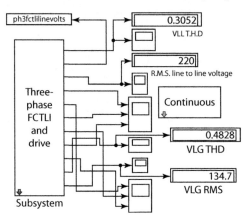

FIGURE 9.10
Three-phase FCTLI model.

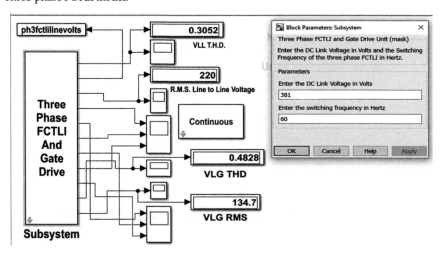

FIGURE 9.11
Three-phase FCTLI model with interactive blocks.

Interactive Models for Three-Phase Multilevel Inverters

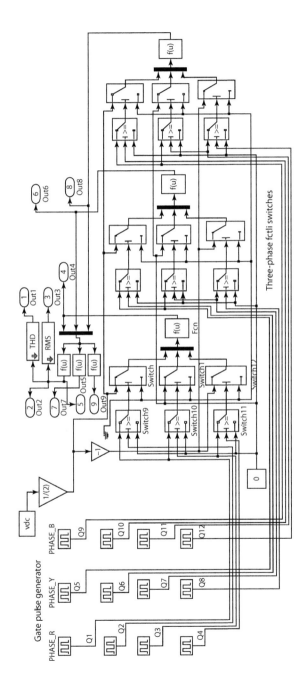

FIGURE 9.12
Three-phase FCTLI model subsytem.

Q5 to Q8 or Q9 to Q12 in respective order. The start point of the appropriate phases may be noted from the remarks column of Table 9.4. If d is the time delay between any two neighbouring or consecutive states in Table 9.4, then this delay d is $\pi/6$ rad. Table 9.4 is implemented as shown in Figure 9.12 using Pulse Generator blocks, marked 'Gate Pulse Generator' in Figure 9.12. The amplitude of all the pulses in all the Pulse Generator blocks is set to 1 and the period is set to (1/fsw), where fsw is the switching frequency of the inverter. The phase delay for each of the switches in Phases R, Y and B is tabulated in Table 9.5. In Table 9.4, gate pulse Q1 is HIGH (logic 1) for nine entries and LOW (logic 0) for three entries. Hence, its duty cycle is 9/12 $((9*\pi/6)/(12*\pi/6))$ or 0.75. The duty cycle for the remaining switches can be calculated in the same way from Table 9.4. Switch names are shown in Figure 9.8.

Three-phase FCTLI switches are implemented using threshold switches and summing Fcn blocks, as shown in Figure 9.12. The switches Q1 to Q4 in Phase R are implemented as per Table 9.4, according to the following modelling statement:

TABLE 9.4

Switching Table for Three-Phase FCTLI

Q1	Q2	Q3	Q4	Vrg	Remarks
1	1	0	0	$+V_{dc}/2$	Phase R start
1	1	0	0	$+V_{dc}/2$	
1	1	0	0	$+V_{dc}/2$	
1	0	1	0	0	
1	0	1	0	0	Phase B start
1	0	1	0	0	
0	0	1	1	$-V_{dc}/2$	
0	0	1	1	$-V_{dc}/2$	
0	0	1	1	$-V_{dc}/2$	Phase Y start
1	0	1	0	0	
1	0	1	0	0	
1	0	1	0	0	
1	1	0	0	$+V_{dc}/2$	Phase R start for next cycle

TABLE 9.5

Three-Phase FCTLI: Gate Pulse Phase Delay

Sl.No.	Switch Phase R	Phase Delay (s)	Switch Phase Y	Phase Delay (s)	Switch Phase B	Phase Delay (s)
1	Q1	9/(12.fsw)	Q5	1/(12.fsw)	Q9	5/(12.fsw)
2	Q2	0	Q6	4/(12.fsw)	Q10	8/(12.fsw)
3	Q3	3/(12.fsw)	Q7	7/(12.fsw)	Q11	11/(12.fsw)
4	Q4	6/(12.fsw)	Q8	10/(12.fsw)	Q12	2/(12.fsw)

Interactive Models for Three-Phase Multilevel Inverters 221

```
If (gate pulse for Q1 is HIGH AND Q2 is HIGH) then
                Output V_rg = +V_dc/2
Else if (gate pulse for Q1 is HIGH AND Q3 is HIGH) then
                Output V_rg = 0
Else if (gate pulse for Q3 is HIGH AND Q4 is HIGH) then
                Output V_rg = -V_dc/2
                End if
```

Similar modelling statements hold good for switches in Phases Y and B. The AND gate is implemented using the threshold switch as shown in Figure 9.5c, which has already been explained in Section 9.2.1. This AND gate principle is used for the three-phase FCTLI switches in Figure 9.12. Referring to the threshold switches from top to bottom on the extreme left of the position marked 'Three-Phase FCTLI Switches' in Figure 9.12, the top switch marked Switch9 performs the AND operation of the gate pulse for Q1 and Q2, the middle switch marked Switch10 performs the AND operation of the gate pulse for Q1 and Q3 and the bottom switch marked Switch11 performs the AND operation of the gate pulse for Q3 and Q4. The top, middle and bottom switch outputs on the extreme left are passed to the $u(2)$ inputs of the second column of the top, middle and bottom switches marked Switch, Switch1 and Switch2, respectively. The $u(1)$ input to the switches marked Switch, Switch1 and Switch2 in the second column are $+V_{dc}/2$, zero and $-V_{dc}/2$, respectively, while their $u(3)$ inputs are all zeros. All these switches have a threshold value of 0.5 and output $u(1)$ when the $u(2)$ input is greater than or equal to the threshold value, else the output is the $u(3)$ input and performs the modelling statement given. The output of the switches marked Switch, Switch1 and Switch2 are added in a function block marked Fcn whose output is the line-to-ground voltage V_{rg}. The switches for Phases Y and B are developed in the same way.

9.3.2 Simulation Results

The simulation results of the line-to-line voltage, line-to-ground voltage, RMS value of the line-to-line voltage and its THD are shown in Figure 9.13 for a three-phase 60 Hz FCTLI using the ode23t (mod. stiff/trapezoidal) solver [9]. The DC link voltage is 381 V. The harmonic spectrum of the line-to-line voltage is shown in Figure 9.14a and b, respectively. Simulation results are tabulated in Table 9.6.

9.4 Three-Phase Cascaded H-Bridge Inverter

A three-phase symmetrical five-level cascade H-bridge (CHB) inverter is shown in Figure 9.15 [10–11]. In each phase, there are two H-bridge inverter cells connected in cascade. A separate DC voltage source is connected for each H-bridge, having a value V_{dc}. The total output voltage of each phase is

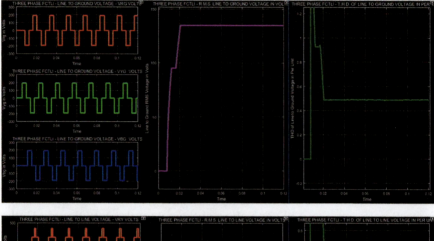
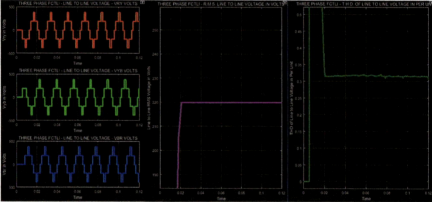

FIGURE 9.13
Three-phase FCTLI simulation results.

the algebraic sum of the voltages across each H-bridge inverter cell. Referring to Phase A, the output voltage v_A can be expressed as follows:

$$v_A = \sum_{i=1,2} v_{ai} \qquad (9.1)$$

Separate DC voltage sources with unequal values can also be connected to each H-bridge inverter cell. Any number of H-bridge inverter units can be cascaded to obtain the required output voltage value, and the output voltage level also increases. Each semiconductor switch in the cascaded H-bridge must be provided with a separate gate drive unit; this makes the design complex. The CHB inverter finds applications in ASDs, electric traction, electric vehicles, battery charging, renewable energy interface, active filter reactive power compensation, flexible AC transmission systems and high-voltage DC (HVDC) transmission.

Referring to the upper H-bridge in Phase A of Figure 9.15, it is seen that the output voltage v_{a2} takes values of $+V_{dc}$ when switches S1A and S4A are ON,

Interactive Models for Three-Phase Multilevel Inverters 223

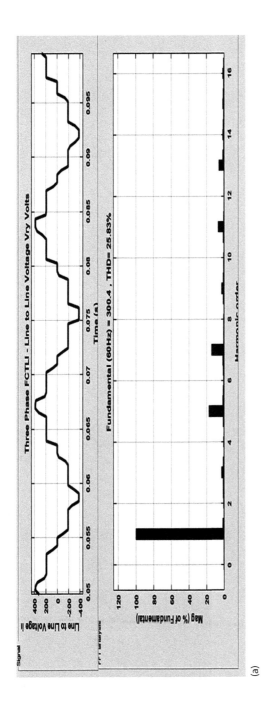

(a)

FIGURE 9.14
(a,b) Three-phase FCTLI – harmonic spectrum of line-to-line voltage V_{ry}.

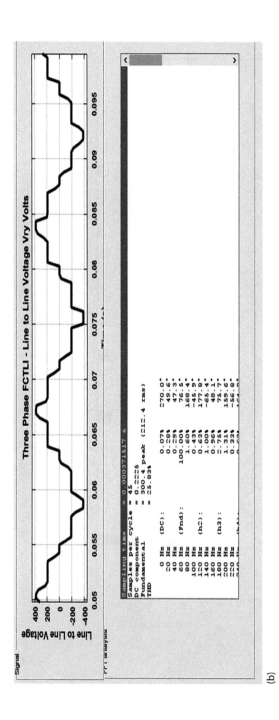

FIGURE 9.14
Three-phase FCTLI. (b) Harmonic spectrum of line-to-line voltage.

Interactive Models for Three-Phase Multilevel Inverters

TABLE 9.6

Three-Phase FCTLI: Simulation Results

Sl. No.	DC Link Voltage (V)	Frequency (Hz)	V_{LL} RMS (V)	V_{LL} THD %	V_{LG} RMS (V)	V_{LG} THD %
1	381	60	220	30.52	134.7	48.28

zero when either S1A and S3A are ON or when S2A and S4A are ON, and $-V_{dc}$ when S2A and S3A are ON. Similar logic holds good for output voltage v_{a1} in the lower H-bridge. Thus, the output phase-to-ground voltage v_A takes values $+2V_{dc}$, $+V_{dc}$, 0, $-V_{dc}$ and $-2V_{dc}$, respectively, giving a five-level voltage. The complete truth table for Phase A of Figure 9.15 is shown in Table 9.7. A similar table is applicable to Phases B and C with switches S1A to S8A replaced by S1B to S8B and S1C to S8C, respectively. The v_A column is replaced by v_B and v_C and the initial phase spread in the 'Remarks' column corresponding to serial no. 1 for phase B starts at 120–165° and that for phase C starts at 240–285°.

The model and simulation of a single-phase five-level cascaded H-bridge inverter using PSIM is shown in Figure 9.16a–c. The gate pulses are derived using Table 9.7. The DC voltage source for each H-bridge is 100 V and the switching frequency is 50 Hz. A series-connected R–L load of 50 Ω and 0.5 H is used. The RMS value of the phase-to-ground voltage is found to be 122.469 V by simulation.

9.4.1 Modelling of Three-Phase Five-Level Cascaded H-Bridge Inverter

The model of the three-phase five-level cascaded H-bridge inverter (TPFLCHBI) is shown in Figure 9.17 with interactive dialogue boxes. The various subsystems are shown in Figure 9.18a–c. A brief description of the various subsystems are given here:

Figure 9.18a corresponds to the dialogue box 'TPCHB Inverter Gate Pulse Generator' shown in Figure 9.17. The switching period of the inverter in seconds is entered in the box provided. The switches S1A to S8A in Phase A, S1B to S8B in Phase B and S1C to S8C in Phase C are marked with their respective outputs. The dialogue box for switch S1A in Phase A is also shown in Figure 9.18a. Referring to Table 9.7, it is seen that neighbouring or consecutive entries differ by a phase delay d, which is 45° ($\pi/4$ rad) or $T/8$ s, where T is the inverter switching period. Referring to the switch S1A column in Table 9.7, it is seen that there are three '1' (ON) and five '0' (OFF) states. Thus, the duty cycle of the gate pulse for S1A is $(3*\pi/4)/(8*\pi/4)$, which is 3/8 or 300/8 per cent. The duty cycle of the gate pulse for the remaining switches can be calculated in the same way. Also, the gate pulse for S1A goes ON (logic 1) at 315° and remains ON until 450°. This gate pulse phase delay is $7*T/8$ s. Pulse amplitude logic 1 means HIGH. These values are entered in the dialogue box for S1A in Figure 9.18a. The phase delay for the gate pulse of the remaining switches are calculated in the same way and are tabulated in Table 9.8.

226 Power Electronic Converters

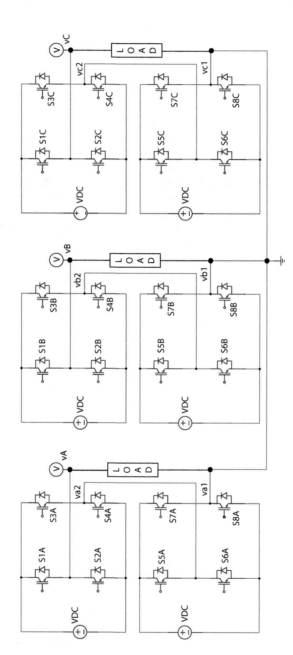

FIGURE 9.15
Three-phase CHB five-level inverter.

Interactive Models for Three-Phase Multilevel Inverters 227

TABLE 9.7

Truth Table: Three-Phase CHB Inverter

Sl. No.	S1A	S2A	S3A	S4A	S5A	S6A	S7A	S8A	v_A (V)	Remarks 1 ON; 0 OFF
1	1	0	0	1	1	0	0	1	$+2V_{dc}$	0°–45°
2	1	0	0	1	0	1	0	1	$+V_{dc}$	45°–90°
3	0	1	0	1	0	1	0	1	0	90°–135°
4	0	1	1	0	0	1	0	1	$-V_{dc}$	135°–180°
5	0	1	1	0	0	1	1	0	$-2V_{dc}$	180°–225°
6	0	1	1	0	0	1	0	1	$-V_{dc}$	225°–270°
7	0	1	0	1	0	1	0	1	0	270°–315°
8	1	0	0	1	0	1	0	1	$+V_{dc}$	315°–360°
9	1	0	0	1	1	0	0	1	$+2V_{dc}$	Next cycle

Figure 9.18b corresponds to the dialogue box 'TPCHB Inverter Model Switches' in Figure 9.17. Here, the DC link voltage for the upper or lower H-bridge is entered in volts and the switching frequency in hertz in the box provided. The model of the semiconductor switches shown in Figure 9.15 is implemented using threshold switches and summing blocks as shown in Figure 9.18b. The switches S1A to S8A in Phase A are implemented as per Table 9.7, according to the following modelling statement:

```
If (gate pulses for S1A AND S4A AND S5A AND S8A are HIGH) then
                 Output V_Ag = +2 * V_dc
Else if (gate pulses for S1A AND S4A AND S6A AND S8A are HIGH) then
                 Output V_Ag = +V_dc
Else if (gate pulses for S2A AND S4A AND S6A AND S8A are HIGH) then
                 Output V_Ag = 0
Else if (gate pulses for S2A AND S3A AND S6A AND S8A are HIGH) then
                 Output V_Ag = -V_dc
Else if (gate pulses for S2A AND S3A AND S6A AND S7A are HIGH) then
                 Output V_Ag = -2 * V_dc
                 End if
```

Similar modelling statements hold good for switches in Phases B and C. The AND gate is implemented using a threshold switch as shown in Figure 9.5c. This is explained in Section 9.2.1. This AND gate principle is used in Figure 9.18b. Referring to threshold switches from top to bottom on the extreme left of Figure 9.18b corresponding to Phase A, the top switch marked Switch1 performs the AND operation of the gate pulse for S1A and S2A, the switch marked Switch2 performs the AND operation of the gate pulse for S5A and S8A and Switch3 performs the AND operation of the output of Switch1 and Switch2. The output of Switch3 is passed to the $u(2)$ input of Switch4, whose u(1) input is $+2*V_{dc}$ and $u(3)$ input is zero. Switch1 to Switch4 have a threshold value of 0.5 and output the $u(1)$ input when $u(2)$ is greater than or equal to 0.5, else output zero value. Thus, when S1A, S4A, S5A and S8A only

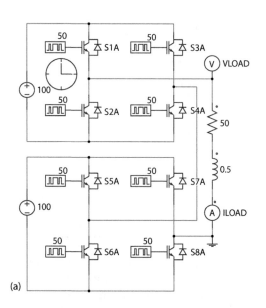

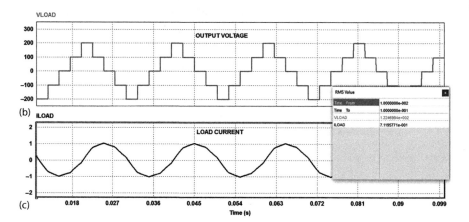

FIGURE 9.16
(a) Single-phase five-level cascaded H-bridge inverter model. (b and c) Single-phase cascaded H-bridge five-level inverter: (b) output voltage; (c) load current.

are ON, the output of Switch4 v_{Ag} is $+2*V_{dc}$ V. All other threshold switches in Phases A, B and C are identical to Switch1 to Switch4. Similarly, Switch5 to Switch8 output is $+V_{dc}$ V when switches S1A, S4A, S6A and S8A only are ON. When S2A, S4A, S6A and S8A only are ON, the output of Switch9 to Switch12 will be 0 V. Swich13 to Switch16 output is $-V_{dc}$ V when S2A, S3A, S6A and S8A only are ON. Finally, Switch17 to Switch20 output is $-2*V_{dc}$ V when S2A, S3A, S6A and S7A only are ON. The final output of the switches marked Switch4, Switch8, Switch12, Switch15 and Switch19 are passed to the Sum of Elements block to obtain the algebraic sum of the output of each of the four threshold

Interactive Models for Three-Phase Multilevel Inverters 229

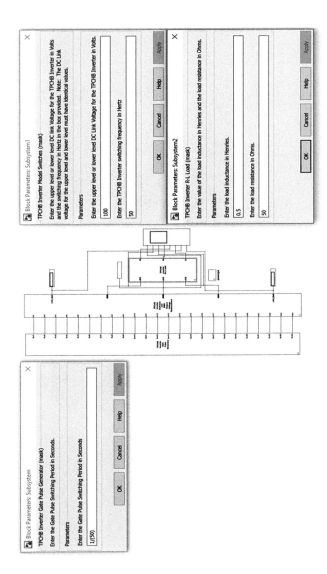

FIGURE 9.17
Model of three-phase five-level CHB inverter with interactive blocks.

230 Power Electronic Converters

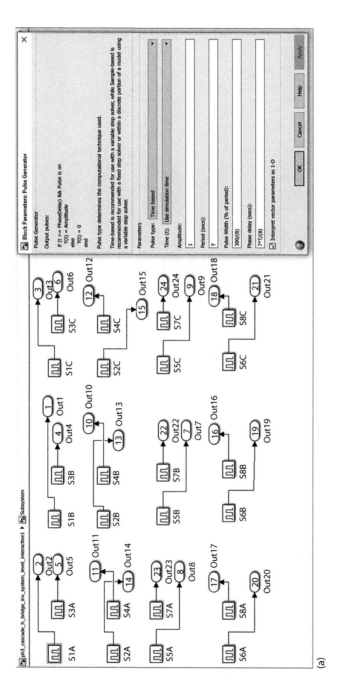

FIGURE 9.18
TPFLCHBI. (a) Gate drive.

Interactive Models for Three-Phase Multilevel Inverters 231

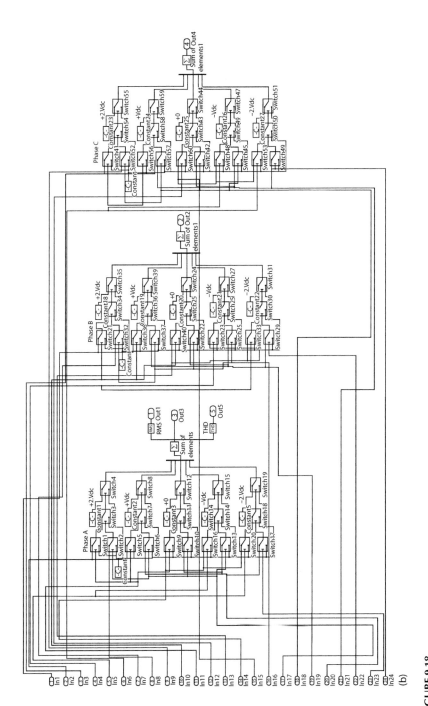

FIGURE 9.18
TPFLCHBI. (b) Switches.

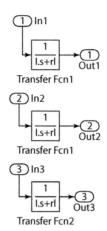

FIGURE 9.18
TPFLCHBI. (c) R-L Load.

TABLE 9.8

TPFLCHBI: Gate Pulse Phase Delay

Sl. No.	Switch Phase A	Phase Delay (s)	Switch Phase B	Phase Delay (s)	Switch Phase C	Phase Delay (s)
1	S1A	7*T/8	S1B	7*T/8−2*T/3	S1C	7*T/8−T/3
2	S2A	2*T/8	S2B	2*T/8−2*T/3	S2C	2*T/8−T/3
3	S3A	3*T/8	S3B	3*T/8−2*T/3	S3C	3*T/8−T/3
4	S4A	6*T/8	S4B	6*T/8−2*T/3	S4C	6*T/8−T/3
5	S5A	0*T/8	S5B	0*T/8−2*T/3	S5C	0*T/8−T/3
6	S6A	T/8	S6B	T/8−2*T/3	S6C	T/8−T/3
7	S7A	4*T/8	S7B	4*T/8−2*T/3	S7C	4*T/8−T/3
8	S8A	5*T/8	S8B	5*T/8−2*T/3	S8C	5*T/8−T/3

switches that represent the output phase voltage v_{Ag}. A similar principle of operation holds good for the threshold switches in Phases B and C.

The three-phase load connected to the TPFLCHBI is a series combination of resistor and inductor. This load model is shown in Figure 9.18c. This load is represented by a Transfer Function block $(1/(ls+rl))$ in each of the three phases.

9.4.2 Simulation Results

The simulation results of the phase-to-ground voltage of the TPFLCHBI is shown in Figure 9.19 for a three-phase 50 Hz inverter switching frequency, using the ode23t (mod. stiff/trapezoidal) solver [9]. The harmonic spectrum of the phase-to-ground voltage of the TPFLCHBI is shown in Figure 9.20. The DC source voltage for each of the H-bridges is 100 V. The RMS value and THD of the phase-to-ground voltage are displayed in Figure 9.19. Simulation results are tabulated in Table 9.9.

Interactive Models for Three-Phase Multilevel Inverters 233

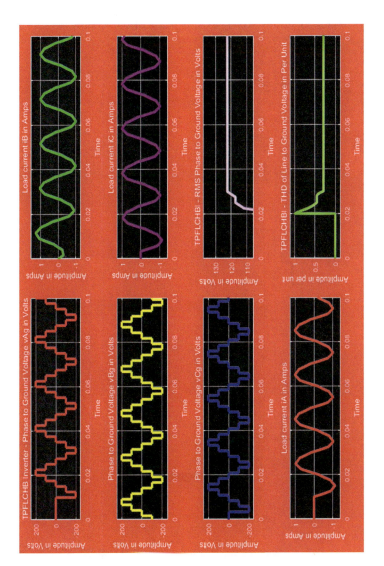

FIGURE 9.19
Simulation results for TPFLCHB inverter.

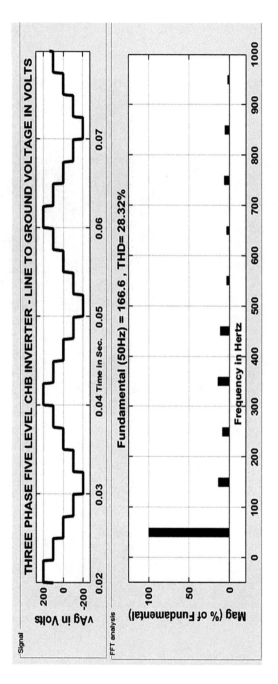

FIGURE 9.20
Harmonic spectrum of line-to-ground voltage of TPFLCHB inverter.

Interactive Models for Three-Phase Multilevel Inverters 235

TABLE 9.9

TPFLCHBI: Simulation Results

Sl. No.	Symmetrical DC Link Voltage Source (V)	Frequency (Hz)	V_{Ag} RMS (V)	V_{Ag} THD (%)
1	100	50	122.5	29.02

9.5 RMS Value and Harmonic Analysis of the Line-to-Line Voltage of Three-Phase DCTLI and FCTLI

The line-to-ground voltage of the three-phase DCTLI and FCTLI is defined as follows:

$$\begin{aligned} V_{rg} &= +\frac{V_{dc}}{2} \quad \text{for} \quad 0 \leq \omega t \leq \pi/2 \\ &= 0 \quad \text{for} \quad \pi/2 \leq \omega t \leq \pi \\ & \quad \text{and} \quad \text{for} \quad 3\pi/2 \leq \omega t \leq 2\pi \\ &= -\frac{V_{dc}}{2} \quad \text{for} \quad \pi \leq \omega t \leq 3\pi/2 \end{aligned} \tag{9.2}$$

$$\begin{aligned} V_{rg} &= +\frac{V_{dc}}{2} \quad \text{for} \quad 2\pi/3 \leq \omega t \leq 7\pi/6 \\ &= 0 \quad \text{for} \quad \pi/6 \leq \omega t \leq 2\pi/3 \\ & \quad \text{and} \quad \text{for} \quad 7\pi/6 \leq \omega t \leq 10\pi/6 \\ &= -\frac{V_{dc}}{2} \quad \text{for} \quad 10\pi/6 \leq \omega t \leq 13\pi/6 \end{aligned} \tag{9.3}$$

Subtracting Equation 9.2 from Equation 9.3, the equation for V_{ry} can be defined as follows:

$$\begin{aligned} V_{ry} &= +V_{dc} \quad \text{for} \quad 0 \leq \omega t \leq \pi/6 \\ &= +\frac{V_{dc}}{2} \quad \text{for} \quad \pi/6 \leq \omega t \leq \pi/2 \\ & \quad \text{and} \quad \text{for} \quad 10\pi/6 \leq \omega t \leq 2\pi \\ &= 0 \quad \text{for} \quad \pi/2 \leq \omega t \leq 2\pi/3 \\ & \quad \text{and} \quad \text{for} \quad 3\pi/2 \leq \omega t \leq 10\pi/6 \\ &= -\frac{V_{dc}}{2} \quad \text{for} \quad 2\pi/3 \leq \omega t \leq \pi \\ & \quad \text{and} \quad \text{for} \quad 7\pi/6 \leq \omega t \leq 3\pi/2 \\ &= -V_{dc} \quad \text{for} \quad \pi \leq \omega t \leq 7\pi/6 \end{aligned} \tag{9.4}$$

The line-to-line voltage harmonics are derived as follows using a Fourier series. Considering one half-cycle of line voltage V_{ry} in Figure 9.7a,b or Figure 9.14a,b and using Equation 9.4, with an axis of symmetry where $f(-t) = -f(t)$, the Fourier expression for the nth harmonic line-to-line voltage V_{ryn} is derived as follows:

$$V_{ryn} = b_n * \sin(n.\omega t) \tag{9.5}$$

The coefficient b_n is derived as follows:

$$b_n = \frac{2}{\pi} * \left[\int_{\pi/12}^{5\pi/12} (V_{dc}/2).\sin(n.\omega t).d(\omega t) + \int_{5\pi/12}^{7\pi/12} V_{dc}.\sin(n.\omega t).d(\omega t) \right.$$
$$\left. + \int_{7\pi/12}^{11\pi/12} (V_{dc}/2).\sin(n.\omega t).d(\omega t) \right] \tag{9.6}$$

Simplifying Equation 9.6, we have the following:

$$b_n = \frac{2V_{dc}.\cos(\pi.n/6)}{\pi.n} * \left[\cos(\pi.n/4) - \cos(3\pi.n/4) \right] \tag{9.7}$$

The RMS value of line-to-line voltage V_{ry} is derived as follows:

$$V_{ry\_rms} = \left[\frac{1}{\pi} * \left[\int_0^{\pi/3} (V_{dc}^2/4)\,d(\omega t) + \int_{\pi/3}^{\pi/2} (V_{dc}^2)\,d(\omega t) + \int_{\pi/2}^{5\pi/6} (V_{dc}^2/4)\,d(\omega t) \right] \right]^{1/2} \tag{9.8}$$

$$V_{ry\_rms} = \frac{V_{dc}}{\sqrt{3}} \tag{9.9}$$

Similarly, referring to the line-to-ground voltage in Figures 9.6 and 9.13 and using Equation 9.2 and noting that each transition from one level to the next are equally spaced at intervals of $\pi/2$ rad, the RMS value of line-to-ground voltage V_{rg} is given by the equation:

$$V_{rg\_rms} = \left[\frac{1}{2\pi} * \left[\int_{\pi/2}^{\pi} (V_{dc}^2/4)\,d(\omega t) + \int_{3\pi/2}^{2\pi} (V_{dc}^2/4)\,d(\omega t) \right] \right]^{1/2} \tag{9.10}$$

$$V_{rg\_rms} = \frac{V_{dc}}{2*\sqrt{2}} \tag{9.11}$$

Interactive Models for Three-Phase Multilevel Inverters

Thus, for a DC link voltage of 381 V, the values of V_{ry_rms} and V_{rg_rms} work out to 220 V and 134.7 V, respectively.

The THD of the line-to-line voltage of a three-phase three-level inverter can be expressed as follows:

$$\text{THD of } V_{ry} = \sqrt{\frac{(V_{dc}/\sqrt{3})^2}{(\sqrt{3}.V_{dc}/\pi)^2} - 1}$$

$$= 30.9\%$$

(9.12)

9.6 RMS Value and THD of Phase-to-Ground Voltage of TPFLCHB Inverter

From Table 9.7, it is clear that each of the five voltage levels of the phase-to-ground voltage v_{Ag} has a duration of $\pi/4$ rad. Referring to Figure 9.16b or Figure 9.20, drawing a vertical axis midway between the line corresponding to 0 V, it is seen that $f(+t) = -f(-t)$ and the Fourier expression for the nth harmonic of v_{Agn} can be expressed as follows:

$$V_{Agn} = b_n * \sin(n.\omega t)$$

(9.13)

Coefficient b_n is derived as follows:

$$b_n = \frac{2}{\pi} * \left[\int_{-\pi/8}^{+\pi/8} 0 * d(\omega t) + \int_{+\pi/8}^{+3\pi/8} V_{dc} * \sin(n.\omega t) * d(\omega t) \right.$$

$$+ \int_{+3\pi/8}^{+5\pi/8} 2.V_{dc} * \sin(n.\omega t) * d(\omega t)$$

(9.14)

$$\left. + \int_{+5\pi/8}^{+7\pi/8} V_{dc} * \sin(n.\omega t) * d(\omega t) + \int_{+7\pi/8}^{+\pi} 0 * d(\omega t) \right]$$

Equation 9.14 simplifies to the following:

$$b_n = \frac{2 * V_{dc}}{n.\pi} * \left[\cos\left(\frac{n.\pi}{8}\right) + \cos\left(\frac{3.n.\pi}{8}\right) - \cos\left(\frac{5.n.\pi}{8}\right) - \cos\left(\frac{7.n.\pi}{8}\right) \right]$$

(9.15)

$$V_{Agn} = \frac{2 * V_{dc}}{n.\pi} * \left[\cos\left(\frac{n.\pi}{8}\right) + \cos\left(\frac{3.n.\pi}{8}\right) - \cos\left(\frac{5.n.\pi}{8}\right) - \cos\left(\frac{7.n.\pi}{8}\right) \right]$$

$$* \sin(n.\omega t)$$

(9.16)

238 *Power Electronic Converters*

Letting n equal 1 in Equation 9.16, the RMS value of the fundamental component is as follows:

$$V_{Ag1\_rms} = 1.1765 * V_{dc} \qquad (9.17)$$

The RMS value of the phase-to-ground voltage V_{Ag_rms} can be derived from Figure 9.16b or Figure 9.20, as follows:

$$V_{Ag\_rms} = \left[\frac{2}{\pi} * \left\{ \int_0^{\pi/4} 0 * d(\omega t) + \int_{\pi/4}^{\pi/2} V_{dc}^2 * d(\omega t) \right. \right.$$

$$\left. \left. + \int_{\pi/2}^{3\pi/4} 4.V_{dc}^2 * d(\omega t) + \int_{3\pi/4}^{\pi} V_{dc}^2 * d(\omega t) \right\} \right]^{\frac{1}{2}} \qquad (9.18)$$

Equation 9.18 simplifies to the following:

$$V_{Ag\_rms} = \sqrt{\frac{3}{2}} * V_{dc} \qquad (9.19)$$

The THD of V_{Ag} is derived as follows:

$$\text{THD } of \ V_{Ag} = \left[\sqrt{\left(\frac{1.2249 * V_{dc}}{1.1765 * V_{dc}} \right)^2 - 1} \right.$$

$$= 0.2898 \text{ per unit} \qquad (9.20)$$

9.7 Pulse Width Modulation Methods for Multilevel Converters

The methods used for PWM of conventional two-level inverters can be extended for multilevel converters [10–11]. The three PWM methods used for multilevel converters can be classified as (1) multilevel carrier-based sine PWM or multi-carrier sine PWM; (2) selective harmonic elimination PWM; and (3) multilevel space vector PWM.

The multi-carrier-based sine PWM method can be classified into two groups: (1) multi-carrier sine phase-shift PWM (MSPSPWM) and (2) multi-carrier sine level shift PWM (MSLSPWM). In this method, depending on the voltage level of the multilevel converter, triangle carriers having the same peak value A_c and carrier frequency f_c Hz are either shifted in phase or shifted in voltage level and then compared with a sine-wave reference signal or modulating signal whose peak-to-peak value is A_m and frequency f_m Hz and is zero centred

Interactive Models for Three-Phase Multilevel Inverters 239

at the middle of the triangle carrier set. The reference sine-wave signal is compared with each of the triangle carrier signals. If the carrier signal is less than the sine reference signal, then the switch corresponding to that carrier is turned ON, and if the carrier signal is greater than the sine reference signal, the relevant switch is turned OFF. In multilevel converters having output voltage levels m, the amplitude modulation and frequency modulation indices m_a and m_f are defined as follows:

$$m_a = \frac{A_m}{(m-1)*A_c} \tag{9.21}$$

$$m_f = \frac{f_c}{f_m} \tag{9.22}$$

In this chapter, interactive system models for these two types of MSPWM are presented. The models developed are suitable for three-phase DCTLI, FCTLI and TPFLCHBI.

9.7.1 Multi-Carrier Sine Phase-Shift PWM

In this method, identical triangle carriers each with a peak value of +1 and −1 V and frequency f_c Hz are shifted in phase from each other by either $360/k$ or $180/k$, where k is the number of triangle carriers used. The value of k is based on the output voltage level of the multilevel converter. For three-phase DCTLI and FCTLI, two triangle carriers are used. For a TPFLCHBI, four triangle carriers are required. The number of triangle carriers required is given by $(m-1)$, where m is the output phase-to-ground voltage levels.

The interactive model of the MSPSPWM generator is shown in Figure 9.21, along with its dialogue boxes. Up to five triangle carriers are provided, suitable for three-phase DCTLI, FCTLI and FLCHBI. The inner subsystems are shown in Figure 9.22a,b. Referring to Figure 9.21, the two subsystems are 'Multi-Triangle Phase Shift Carrier Generator' and 'Multi-Carrier Phase Shift PWM Three-Phase Sine Modulating Signal Generator'. The schematic of the former and latter subsystems are shown in Figure 9.22a,b.

Program Segment 9.1

```
function [va,vb,vc] = fcn(vm,f,t)
va = vm*sin(2*pi*f*t);
vb = vm*sin(2*pi*f*t - 2*pi/(3));
vc = vm*sin(2*pi*f*t + 2*pi/(3));
```

Figure 9.22a generates phase shifted triangle carriers each with a positive and negative peak of +1 and −1 V, respectively. The 'Pulse Generator'

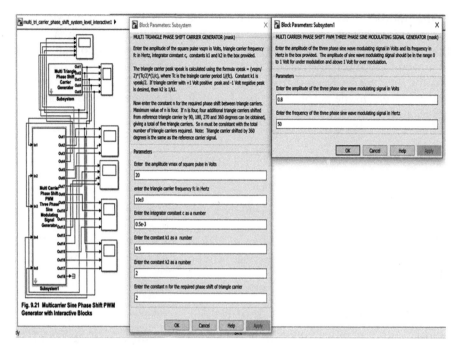

FIGURE 9.21
MCSPSPWM generator with interactive blocks.

and 'Transport Delay1' dialogue boxes are shown in Figure 9.22a. The pulse amplitude entered is 'vsqm' and the frequency entered in hertz is 'fc', which corresponds to the triangle carrier frequency. Inputs to the mux are pulse amplitude vsqm and constant vsqm/2. These two values are subtracted using the Fcn6 block and the resulting output is passed to Transfer Function block $1/c.s$, where c is a constant. The output of the Transfer Fcn block is a triangle carrier with frequency f_c Hz, peak value v_{peak} and minimum value zero. This triangle carrier is the input u to the Fcn7 block that performs the operation $(u-k1)*k2$, where $k1 = v_{peak}/2$ and $k2 = 1/k1$. The output of the Fcn7 block is the triangle carrier having frequency f_c Hz, positive peak +1 V and negative peak −1 V. The constant c is calculated using the formula $v_{peak} = (vsqm/2)*(T_c/2)*1/c$, where $T_c = 1/f_c$ is the carrier period.

The Transport Delay1 to 4 blocks generate the necessary phase shift from the reference triangle carrier. The dialogue box for the Transport Delay1 block is shown in Figure 9.22a. Here, Time Delay is entered as '1/(n*fc)', where n is an integer constant, depending on the number of phase shifted triangle carriers required. The value 1 for n creates no phase shift. The values $n = 2,3,4$ create a total of two, three and four triangle carriers, each phase shifted by 180°, 120° and 90°, respectively.

Interactive Models for Three-Phase Multilevel Inverters 241

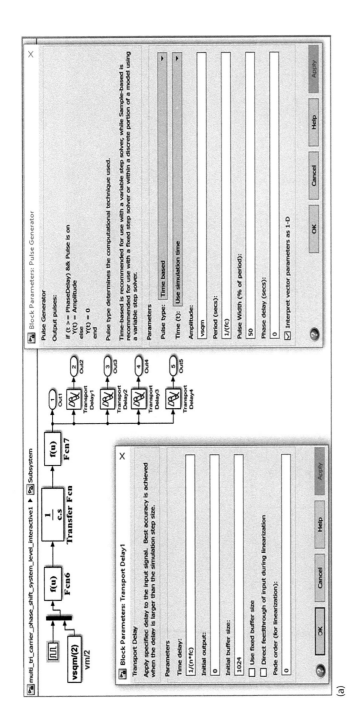

FIGURE 9.22
Multi-carrier (a) generator with interactive blocks.

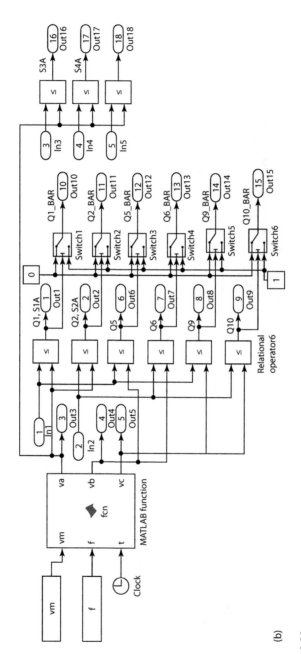

FIGURE 9.22
Multi-carrier (b) sine PWM generator.

Figure 9.22b shows the model of a multi-carrier sine PWM generator, whose dialogue box is shown in Figure 9.21. The MATLAB® function generates a three-phase sine-wave modulating signal with peak value 'vm' (0.8 V) and frequency 'f' (50 Hz) as per the code given in Program Segment 9.1. The respective triangle carrier and the sine-wave modulating signals are compared in Relational Operator blocks to obtain the gate pulses for the multilevel converter switches. For three-phase DCTLI and FCTLI, these gate pulses are also inverted using a NOT gate marked Switch1 to Switch6 in Figure 9.22b. For DCTLI, Q3g = (~Q1g) and Q2g = (~Q4g), where the additional suffix g stands for the gate pulse of the respective switch and the tilde symbol (~) stands for a logic NOT operation. Similarly, for FCTLI, Q4g = (~Q1g) and Q3g = (~Q2g). For TPFLCHBI, these NOT gate outputs are not required. The operation of Switch1 to Switch6 as NOT gates is explained next.

Referring to Switch1 to Switch6 in Figure 9.22b, it is seen that their first input $u(1)$ and third input $u(3)$ are 0 and 1, while the control input $u(2)$ is the gate pulse from the respective Relational Operator block. All the switches from 1 to 6 have a threshold value of 0.5 and output $u(1)$ when $u(2)$ is greater than or equal to 0.5 and output $u(3)$ when $u(2)$ is less than 0.5. Thus, when the gate pulse to the respective Switch1 to Switch6 is HIGH (logic 1), the output is 0 (LOW), and when this gate pulse is LOW (logic 0), the output is 1 (HIGH).

9.7.2 Simulation Results

The model simulation of the MSPSPWM generator was carried out using the ode23t (mod. stiff/trapezoidal) solver [9]. The sine-wave peak amplitude and frequency used are 0.8 and 50°Hz respectively. The triangle carrier frequency is 10°kHz. The simulation results for the gate pulse in Phase R of three-phase DCTLI and FCTLI are shown in Figure 9.23a and b and for three-phase FLCHBI are shown in Figure 9.23c and d.

9.7.3 Multi-Carrier Sine Level Shift PWM

In the multi-carrier sine level shift PWM (MSLSPWM) method, identical triangle carriers with the same frequency (f_c Hz) and amplitude +1 V are shifted in level. These triangle carriers occupy positions symmetrically above and below relative to the x-axis or zero-level axis. The sine-wave modulating signal is centred along the zero-level axis.

The model of the MSLSPWM generator is shown in Figure 9.24a–d. Each one generates a different type of carrier pattern for MSLSPWM. The same model with interactive blocks is shown in Figure 9.25. The internal subsystems of the model are shown in Figure 9.26a,b.

The dialogue boxes relating to subsystems 'Multi-Triangle Carrier PWM Generator' and 'Multi-Carrier Level Shift Three-Phase Sine PWM Generator' are shown in Figure 9.25. The method of generating the level-shifted triangle

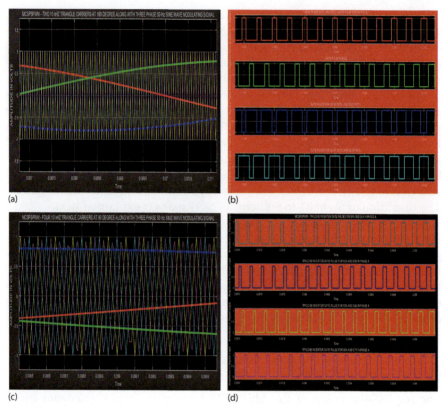

FIGURE 9.23
MCSPSPWM. (a) Two 10 kHz triangle carriers and three-phase 50 Hz sine-wave modulating signal. (b) Gate pulse for Phase R of three-phase DCTLI and FCTLI. (c) Four 10 kHz triangle carriers and three-phase 50 Hz sine-wave modulating signal.

carrier is shown in Figure 9.26a and the MSLSPWM three-phase sine PWM gate pulse generator block is shown in Figure 9.26b. In Figure 9.25, in the dialogue box 'Multi-Triangle Carrier PWM Generator', the relevant data relating to the peak value of the square pulse v_{sqm}, carrier frequency f_c Hz and integrator constant c and the peak value of the triangle carrier are entered. In the dialogue box 'Multi-Carrier Level Shift Three-Phase Sine PWM Generator', the data relating to the amplitude and frequency of the modulating sine wave are entered.

Figure 9.26a shows the model to generate the triangle carrier with a peak of +1 V, minimum value zero and frequency f_c Hz. The Pulse Generator and Transport Delay dialogue boxes are shown in Figure 9.26a. The pulse amplitude entered is 'vsqm' and the frequency entered in hertz is 'fc', which corresponds to the triangle carrier frequency. Inputs to the mux are pulse amplitude vsqm and constant vsqm/2. These two values are subtracted using

Interactive Models for Three-Phase Multilevel Inverters 245

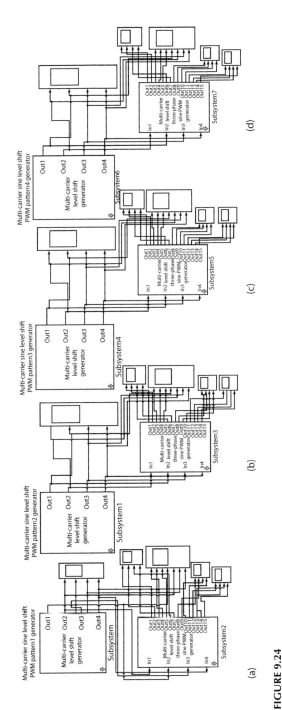

FIGURE 9.24
(a–d) MCSLSPWM gate pulse generator.

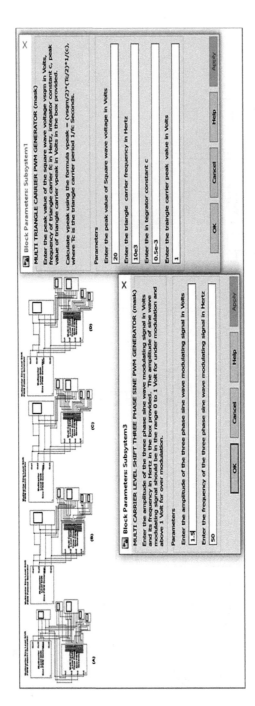

FIGURE 9.25
MCSLSPWM gate pulse generator with interactive blocks.

Interactive Models for Three-Phase Multilevel Inverters 247

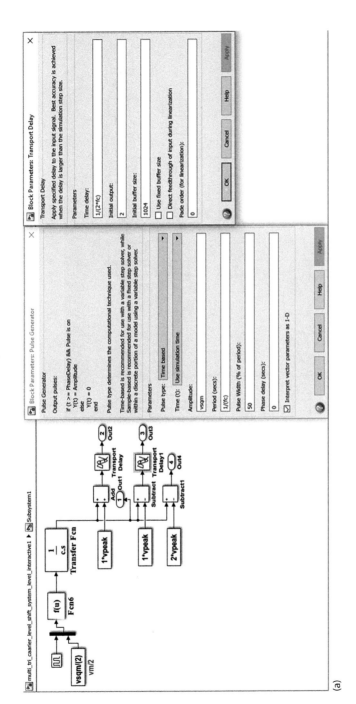

FIGURE 9.26
MCSLSPWM. (a) Carrier generator.

248 · Power Electronic Converters

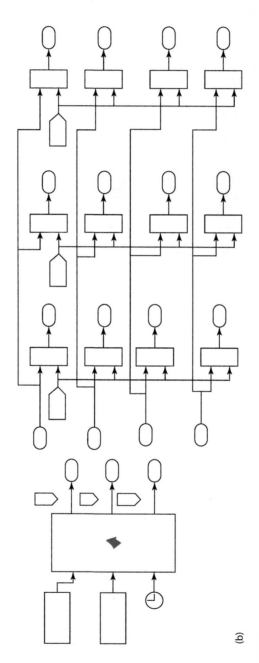

(b)

FIGURE 9.26
MCSLSPWM. (b) Gate pulse generator.

Interactive Models for Three-Phase Multilevel Inverters 249

the Fcn6 block and the resulting output is passed to the Transfer Function block $1/c.s$, where c is a constant. The output of the Transfer Fcn block is a triangle carrier with frequency f_c Hz, peak value v_{peak} and minimum value zero. The constant c is calculated using the formula $v_{peak} = (vsqm/2) * (T_c/2) * (1/c)$, where $T_c = 1/f_c$ is the carrier period. The output of the Transfer Fcn block is a triangle carrier with peak value +1 V, minimum value zero and frequency f_c Hz.

In the Transport Delay and Transport Delay1 dialogue boxes, the time delay entered is '1/(2*fc)'. The triangle carrier output of Transfer Fcn block is added with $1 * v_{peak}$, subtracted from $1 * v_{peak}$ and $2 * v_{peak}$ and delayed by $1/(2 * f_c)$ with two Transport Delay blocks to obtain four different triangle carrier patterns. Figure 9.26a corresponds to carrier pattern B in Figure 9.24. The carrier pattern in Figure 9.24a has *no* Transport Delay blocks. Carrier patterns C and D in Figure 9.24 are generated by relocating the position of the Transport Delay and Transport Delay1 blocks in Figure 9.26a. Referring to Figure 9.26a, by relocating the Transport Delay and Transport Delay1 blocks to the output of the Add and Subtract1 blocks generates one additional carrier pattern, and relocating them to the output of the Subtract and Subtract1 blocks generates a second additional pattern.

Figure 9.26b shows the model of the MSLSPWM gate pulse generator for the switches of a multilevel converter, using the MATLAB Function and Relational Operator blocks. This model is the same as that already explained in Section 9.7.1.

9.7.4 Simulation Results

The simulation of the model of the MSLSPWM generator was carried out using the ode23t (mod. stiff/trapezoidal) solver [9]. The simulation results for the triangle carrier with a three-phase modulating sine-wave and gate pulse for Phase A of the TPFLCHBI are shown in Figure 9.27a–h. The triangle carrier frequency is 10 kHz. The three-phase sine-wave modulating signal has an amplitude of 1.5 V and frequency of 50 Hz. For three-phase DCTLI and FCTLI, two carriers above and below the x-axis with the sine-wave peak below +1 V along with NOT gates have to be used, as explained in Section 9.7.1.

9.8 Conclusions

The system models for three-phase DCTLI, FCTLI and FLCHBI have been successfully developed. In these three models, the time intervals for all the voltage levels are equal. By reducing the time duration for the zero voltage level to a lower value compared with the other two voltage levels, it is

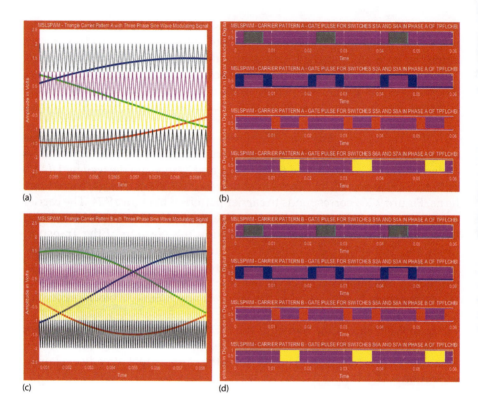

FIGURE 9.27
(a) Triangle carrier pattern A with three-phase sine-wave modulating signal (left). (b) Phase A gate pulse for switches of TPFLCHBI (right). (c) Triangle carrier pattern B with three-phase sine-wave modulating signal (left). (d) Phase A gate pulse for switches of TPFLCHBI (right).

possible to reduce the THD of the line-to-line voltage and improve its harmonic spectrum still further [5]. The THD of the line-to-line voltage for the conventional three-phase three-level inverter with equal time duration for all three voltage levels is 31.08% [5]. The THD of the line-to-line voltage by simulation for this DCTLI is found to be around 30.82% and that for the FCTLI is found to be 30.52%, respectively, whereas the calculated value is 30.9%. The RMS line-to-line voltage for the DCTLI and for the FCTLI agree well with the formula given by Equations 9.7 and 9.8. The THD of the phase-to-ground voltage of the TPFLCHBI is found to be 28.98%, whereas by model simulation this value is 29.02%. Similarly, the RMS value of the phase-to-ground voltage of the TPFLCHBI is found to be 122.49 V by derivation, whereas this value is 122.5 V by model simulation. It is also seen from Figures 9.1, 9.8 and 9.15 that the voltage stress across individual semiconductor switches is reduced for a given DC link voltage for the former two topologies and for a given output

Interactive Models for Three-Phase Multilevel Inverters 251

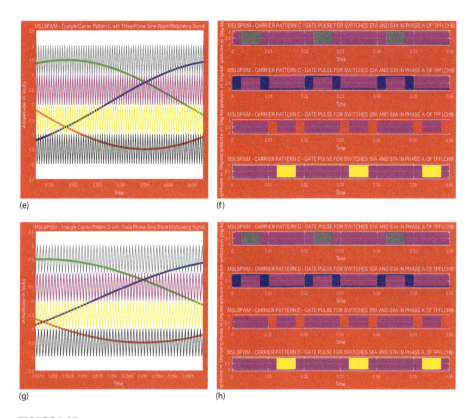

FIGURE 9.27
(e) Triangle carrier pattern C with three-phase sine-wave modulating signal (left). (f) Phase A gate pulses for switches of TPFLCHBI (right). (g) Triangle pattern D with three-phase sine-wave modulating signal (left). (h) Gate pulse for Phase A switches of TPFLCHBI (right).

voltage for the third topology, as compared with the conventional three-phase two-level inverter. As the multilevel triangle carrier PWM technique is used for voltage control of multilevel converters, a section on the development of interactive models for MSPSPWM and MSLSPWM techniques are presented with simulation results.

References

1. C. Hochgraf, R. Lasseter, D. Divan, and T.A. Lipo: "Comparison of multilevel inverters for static VAR compensation", *IEEE*; Industry Applications Society Annual Meeting, Denver, CO, 1994; pp. 921–928.

2. J.S. Lai and F.Z. Peng: "Multilevel convertors: A new breed of power convertors", *IEEE Transactions on Industry Applications*; Vol.32, No.3, May/June 1996; pp. 509–517.
3. F.Z. Peng, J.S. Lai, J.W. Mckeever and J. Van Coevering: "A multilevel voltage source inverter with separate DC sources for static var generation", *IEEE Transactions Industry Applications*; Vol. 32, No. 5, September/October 1996; pp. 1130–1133.
4. J. Rodriguez, J.S. Lai and F.Z. Peng: "Multilevel inverters: A survey of topologies control and applications", *IEEE Transactions on Industrial Electronics*; Vol.49, No.4, August 2002; pp. 724–738.
5. P.M. Bhagwat and V.R. Stefanovic: "Generalized structure of a multilevel PWM inverter", *IEEE Transactions on Industry Applications*; Vol.IA-19, No.6, November/December 1983; pp. 1057–1069.
6. A. Nabae, I. Takahashi, and H. Akagi: "A neutral-point clamped PWM inverter", *IEEE Transactions on Industry Applications*; Vol.IA-17, No.5, September/October 1981; pp. 518–523.
7. Spectrum Software: MICROCAP 11, Demo version, 2016.
8. The Mathworks Inc.: MATLAB/Simulink Release Notes, R2016b, 2016.
9. N.P.R. Iyer: "MATLAB/Simulink modules for modelling and simulation of power electronic converters and electric drives", M.E. (research) thesis, University of Technology Sydney, NSW, Australia, Chapter 12, 2006.
10. J.W. Dixon: "Multilevel converters" in *Power Electronic Converters and Systems Frontiers and Applications*, Ed. A.M. Trzynadlowski, London: The Institution of Engineering and Technology, 2016, pp. 43–72.
11. S. Khomfoi and L.M. Tolbert: "Multilevel power converters" in *Power Electronics Handbook*, Ed. M.H. Rashid, Burlington, MA: Elsevier, Butterworth-Heinemann, 2011, pp. 455–484.

10

Interactive Model Verification

10.1 Introduction

In this chapter, interactive component-level models for the power electronic converters discussed in the previous chapters are presented. These are developed using active and passive components used in the power electronic circuits such as semiconductor switches, diodes, the AC source, the DC source and passive components such as resistors, inductors and capacitors. These active and passive components are mainly from the Power Systems block set in Simulink® [1]. The component-level model simulation results presented here provide verification of the system model simulation results of power electronic converters discussed in the previous chapters.

10.2 AC to DC Converters

In this section, interactive component-level models for the single-phase full-wave diode bridge rectifier (SFWDBR), single-phase full-wave silicon-controlled rectifier (SCR) converter and three-phase full-wave diode bridge rectifier (TFWDBR) discussed in Chapter 3 are presented [2–5].

10.2.1 Single-Phase Full-Wave Diode Bridge Rectifier

The SFWDBR system model presented in Section 3.1 is verified in this section using interactive component-level models [2–5]. The data shown in Table 3.1 are used here. This model with interactive blocks is shown in Figure 10.1. The inner subcircuits of SFWDBR consisting of a semiconductor diode bridge and the purely resistive/RLE load is shown in Figure 10.2a,b. Simulation results for the SFWDBR are shown in Figure 10.3 for RLE load and tabulated in Table 10.1, closely agreeing with Table 3.2.

Simulation results for SFWDBR with purely resistive load can be found similarly and are not shown here.

253

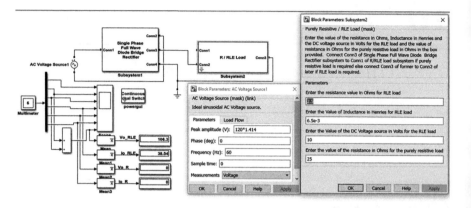

FIGURE 10.1
Model of SFWDBR with interactive blocks.

10.2.2 Single-Phase Full-Wave SCR Bridge

The circuit schematic of the single-phase full-wave controlled bridge rectifier (SFWCBR) is shown in Figure 3.7 and the derivations for output voltage and current for purely resistive and RLE loads are shown in Section 3.2 [2–5]. In this section, a component-level model for SFWCBR is developed and simulation is carried out for an RLE load as per data given in Table 3.3 and compared.

This model of the SFWCBR is shown in Figure 10.4 and interactive blocks are shown in Figure 10.5. The subsystems for this model are shown Figure 10.6a–e. Figure 10.6a is the single-phase full-wave thyristor bridge rectifier, Figure 10.6b is the thyristor gate firing pulse generator using the Pulse Generator block, Figure 10.6c is the R/RLE load, Figure 10.6d selects either the R or RLE load and Figure 10.6e is the output voltage and current measurement unit.

Referring to Figure 10.6d, if the constant entered is zero, then the switch whose threshold value is 0.5 outputs −1 and the logic comparator on the top marked '<0' becomes HIGH and turns on S2, selecting the RLE load. If the constant entered is 1, then the bottom logic comparator marked '>0' output becomes HIGH, turns on S1 and selects the purely resistive load.

The simulation results for the RLE load are shown in Figure 10.7. The results are tabulated in Table 10.2. The simulation results for the purely resistive load are not shown here. The simulation results in Table 10.2 agree closely with those shown in Table 3.4.

10.2.3 Three-Phase Full-Wave Diode Bridge Rectifier

The TFWDBR discussed in Section 3.3 is presented here with a component-level model. The circuit configuration of the TFWDBR is shown in Figure 3.13 [2–5]. The model parameters are shown in Table 3.5. This model of the

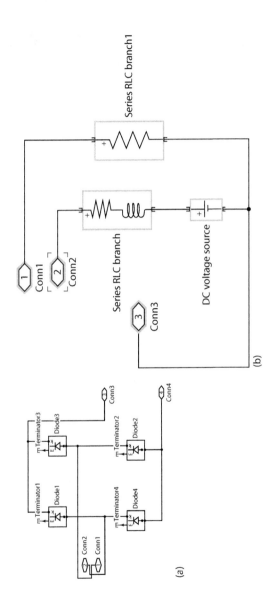

FIGURE 10.2
Subcircuits of SFWDBR. (a) Full bridge diode rectifier. (b) RLE and purely resistive loads.

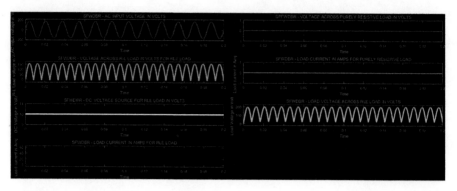

FIGURE 10.3
SFWDBR: simulation results for RLE load.

TABLE 10.1

SFWDBR: Simulation Results for RLE Load

Sl. No.	Type of Load	Udc(RMS) (V)	Udc(AVG) (V)	I_load(AVG) (A)
1	RLE	120	106.3	38.5

TFWDBR is shown in Figure 10.8. The interactive blocks of this model and its inner subsystems are shown in Figure 10.9 and Figure 10.10a–c, respectively. The three-phase diode rectifier is a built-in model from the Power System block set. This is marked 'Rectifier' in Figure 10.8. The three-phase AC voltage source, resistive load and measurement blocks are shown in Figure 10.10a–c, respectively. The simulation results of this model are shown in Figure 10.11a,b, respectively. The results are tabulated in Table 10.3. The simulation results in Table 10.3 agree closely with those shown in Table 3.6.

10.3 DC to AC Converters

In this section, interactive component-level models are developed using semiconductor and passive components for three-phase 180° mode and 120° mode inverters discussed in Chapter 4 [2–5]. The three-phase sine pulse width modulation (PWM) inverter component-level model is presented in Section 10.9. The results obtained by component-level simulation are compared with those in Chapter 4.

Interactive Model Verification

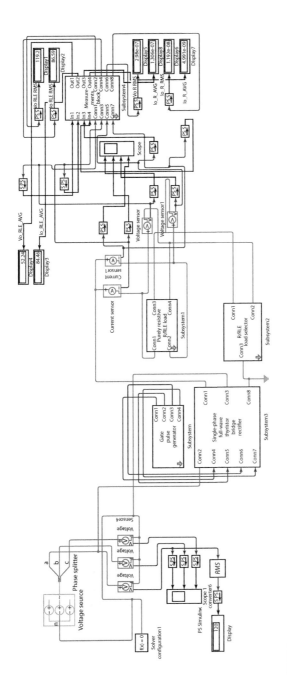

FIGURE 10.4
SFWCBR model.

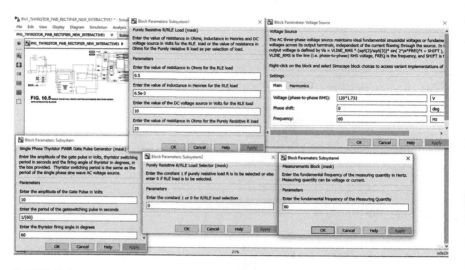

FIGURE 10.5
SFWCBR model with interactive blocks.

10.3.1 Three-Phase 180° Mode Inverter

Detailed analysis of the three-phase 180° mode inverter is given in Section 4.2. The system model for the three-phase 180° mode inverter is presented in Section 4.2.4. Here, the interactive component-level model for this inverter is developed using semiconductor and passive components in the Power Systems block set; the simulation is carried out and compared with that presented in Chapter 4.

This model of the three-phase 180° mode inverter is shown in Figure 10.12, the interactive blocks are shown in Figure 10.13 and the inner subsystems are shown in Figure 10.14a–c [2–5]. Simulation results are shown in Figure 10.15a–d and tabulated in Table 10.4.

Tabulated results for the line-to-line voltage agree well with Tables 4.1 and 4.2.

10.3.2 Three-Phase 120° Mode Inverter

Detailed analysis of the three-phase 120° mode inverter is presented in Section 4.3 A detailed system model for this inverter is also presented, in Section 4.3.4. In this section, an interactive component-level model for this inverter is presented; a simulation was carried out and compared with that given in Section 4.3.

This model of this inverter is the same as for the 180° mode inverter shown in Figures 10.12, 10.13 and 10.14. In this 120° mode, the gate pulse duration is only for $(T/3)$ seconds, where T is the switching period of the inverter [2–5]. Referring to Figure 10.13, in the Gate Pulse Generator block, against the box 'Enter the Duty-cycle of gate pulse as a percentage', the value 33.3333

Interactive Model Verification

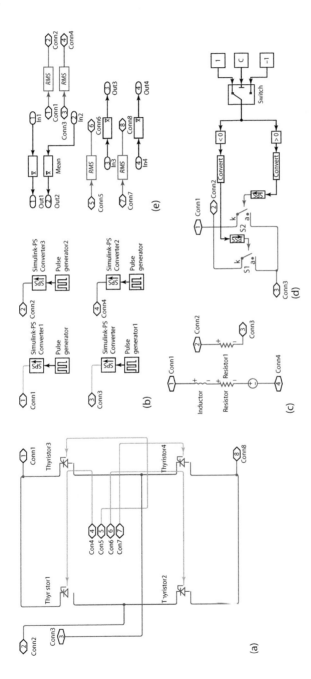

FIGURE 10.6
(a–e) SFWCBR subsystems.

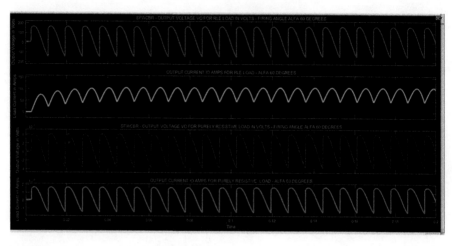

FIGURE 10.7
SFWCBR simulation results for RLE load.

TABLE 10.2

SFWCBR: Simulation Results with RLE Load: Firing Angle $\alpha = 60°$

Sl. No.	Type of Load	V_{dc}(RMS) (V)	V_{dc}(AVG) (V)	I_load(RMS) (A)	I_load(AVG) (A)	P_out (W)
1	RLE	119.2	52.24	86.59	84.46	4412.19

is entered. The same three-phase inverter then performs as a 120° mode inverter.

A simulation of this inverter was carried out and the results are presented in Figure 10.16a–d. The results are also tabulated in Table 10.5. The tabulated results in Table 10.5 agree well with the results tabulated in Tables 4.3 and 4.4.

10.4 DC to DC Converter

In this section, interactive component-level models for buck, boost and buck–boost converters are developed using semiconductor and passive components [2–5,6,7]. This model simulation is performed for these DC to DC converters and the results obtained are compared with those presented in Chapter 5.

10.4.1 Buck Converter

The buck converter analysis in the continuous conduction mode (CCM) and in the discontinuous conduction mode (DCM) is given in Sections 5.2 and 5.3. The data given in Table 5.1 are used here for circuit simulation of the buck converter. The component-level model of the buck converter

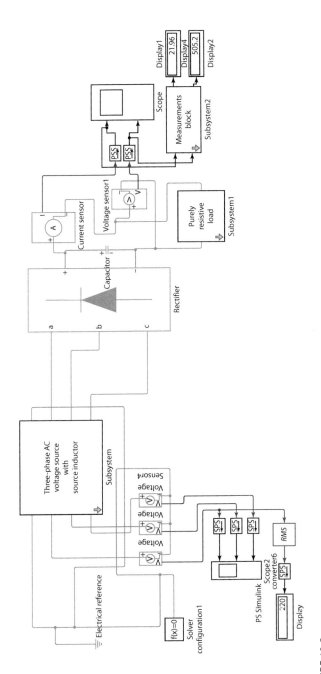

FIGURE 10.8
TFWDBR model.

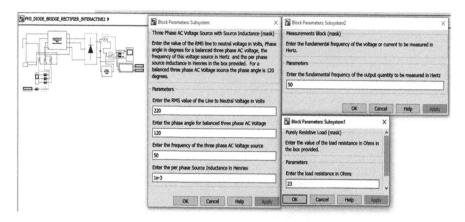

FIGURE 10.9
TFWDBR model with interactive blocks.

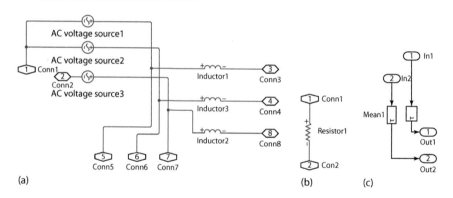

FIGURE 10.10
(a) Three-phase AC voltage source with source inductance. (b) R load. (c) Measurements unit.

is shown in Figure 10.17 with interactive blocks [6,7]. The subsystems are shown in Figure 10.18a,b. The simulation of this buck converter is carried out for the parameters shown in Table 5.1. The simulation results are shown in Figures 10.19 and 10.20 for CCM and DCM, respectively. The simulation results for both cases are tabulated in Table 10.5. Comparison of the results shown in Table 10.5 with Tables 5.2 and 5.3 reveals that they almost agree. The discrepancies are minor. In this model, the semiconductor switch has a forward voltage drop and on resistance, whereas in the system model of the buck converter, the switch is ideal.

10.4.2 Boost Converter

The analysis of the boost converter in CCM and DCM is shown in Sections 5.5 and 5.6. The data shown in Table 5.4 are used here for this model. Component-level model simulation was carried out for both CCM and DCM

Interactive Model Verification 263

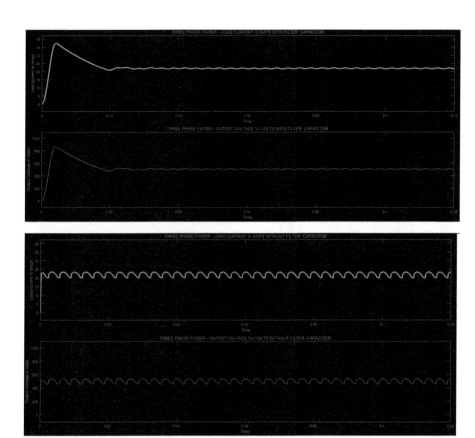

FIGURE 10.11
TFWDBR: simulation results. (a) With capacitor filter. (b) Without filter capacitor.

TABLE 10.3

Three-Phase FWDBR Simulation Results

Sl. No.	Average Load Voltage (V)	Average Load Current (A)	Remarks
1	506.7	22.03	No filter capacitor
2	505.4	21.97	With filter capacitor of 0.0011 F

and the simulation results obtained are compared with those presented in Section 5.7.1.

This model of the boost converter with interactive blocks using semiconductor switches and passive components is shown in Figure 10.21, and the subsystems of this model are shown in Figure 10.22 [6,7]. Simulation results of this model for CCM and DCM operation are shown in Figures 10.23 and 10.24, respectively. Simulation results are tabulated in Table 10.6.

264 *Power Electronic Converters*

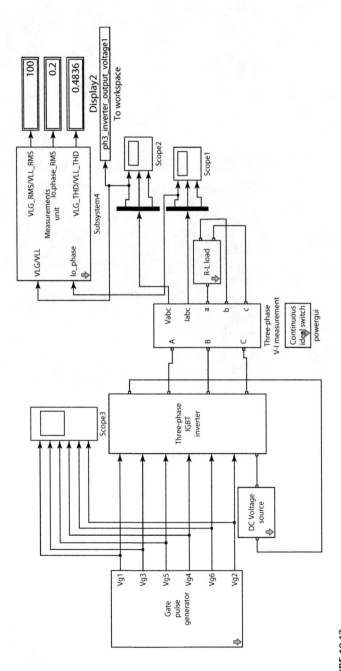

FIGURE 10.12
Three-phase inverter model.

Interactive Model Verification

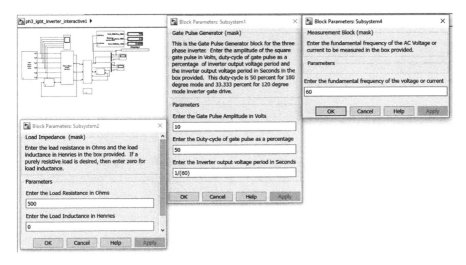

FIGURE 10.13
Three-phase inverter model with interactive blocks.

The simulation results in Table 10.6 almost agree with the results presented in Tables 5.5 and 5.6. In this model, the semiconductor components have forward voltage and on resistance and are not ideal.

10.4.3 Buck–Boost Converter

Analysis of the buck–boost converter in CCM and DCM is shown in Sections 5.8 and 5.9. The data shown in Table 5.7 are used here for the circuit model. A component-level model simulation was carried out for both CCM and DCM and the simulation results obtained are compared with those presented in Section 5.10.1.

This model of the buck–boost converter with interactive blocks is shown in Figure 10.25, and the model subsystems are shown in Figure 10.26 [6,7]. The simulation results of this model for CCM and DCM operation are shown in Figures 10.27 and 10.28. The results are tabulated in Table 10.7. Comparison of the circuit model simulation results in Table 10.7 with Tables 5.8 and 5.9 reveals that the results agree closely. As mentioned, the semiconductor components in the circuit model are not ideal.

10.5 AC to AC Converter

The component-level models for the two three-phase AC controllers discussed in Chapter 6 is presented in this section. The first one is the three-phase back-to-back connected thyristor AC to AC controller connected

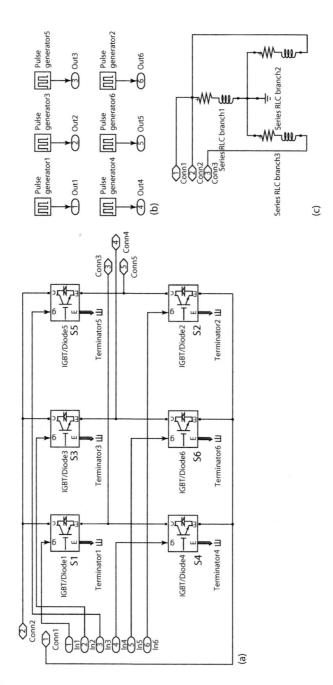

FIGURE 10.14
Three-phase 180° mode inverter model subsystems.

Interactive Model Verification

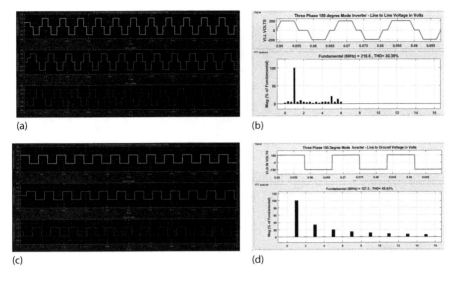

FIGURE 10.15
Three-phase 180° mode inverter. (a) Line-to-line voltages. (b) Line-to-line voltage harmonic spectrum. (c) Line-to-ground voltage. (d) Line-to-ground voltage harmonic spectrum.

to a star-connected resistive load with isolated neutral. The second one is the three-phase back-to-back connected thyristor AC controller in series with resistive load in delta [2–5,8].

10.5.1 Three-Phase Thyristor AC to AC Controller Connected to Resistive Load in Star

The component-level model of the back-to-back connected thyristor AC to AC controller in series with star-connected resistive load is shown in Figure 10.29 [2–5,8]. The interactive blocks of this circuit model and its subsystems are shown in Figures 10.30 and 10.31, respectively. The model parameters are shown in Table 6.1. The model simulation results are shown in Figures 10.32a,b and 10.33a,b for a firing angle α of $\pi/6$ and $2\pi/3$ rad. The

TABLE 10.4

Three-Phase (a) 180° and (b) 120° Mode Inverter: Simulation Results

Sl. No.	Frequency (Hz)	DC Link Voltage (V)	THD of V_{LL} (Per Unit)	RMS of V_{LL} (V)	THD of V_{LG} (Per Unit)	RMS of V_{LG} (V)
(a)						
1	60	200	0.3115	163.3	0.45	100
(b)						
1	60	200	0.3115	141.4	0.3115	81.65

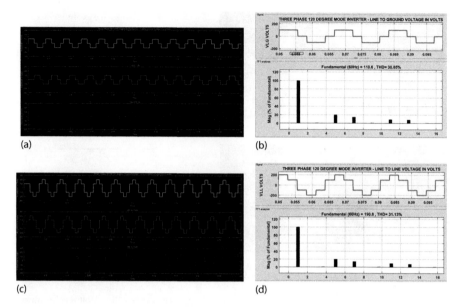

FIGURE 10.16
Three-phase 120° mode inverter. (a) Line-to-ground voltage. (b) Line-to-ground voltage harmonic spectrum. (c) Line-to-line voltage. (d) Line-to-line voltage harmonic spectrum.

TABLE 10.5

Buck Converter Simulation Results

Sl. No.	Output Voltage V_O (V)	Minimum Inductor Current (A)	Maximum Inductor Current (A)	Load Current (A)	Remarks
1	4.74	0.343	3.516	1.896	CCM
2	7.112	0	8.159	2.845	DCM

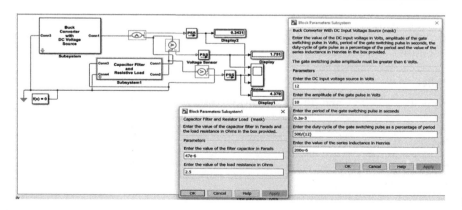

FIGURE 10.17
Buck converter model.

Interactive Model Verification 269

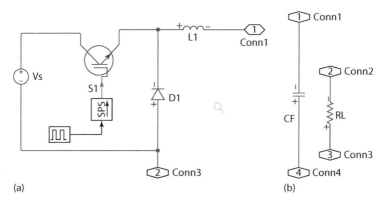

FIGURE 10.18
Buck converter model subsystems. (a) Buck converter. (b) Capacitor filter and resistance load.

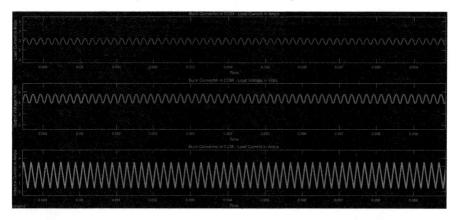

FIGURE 10.19
Buck converter in CCM: simulation results.

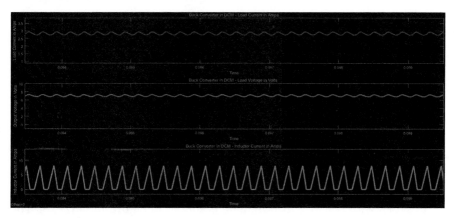

FIGURE 10.20
Buck converter in DCM: simulation results.

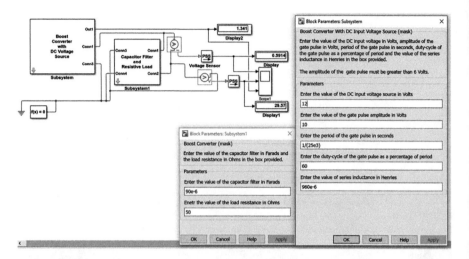

FIGURE 10.21
Boost converter model.

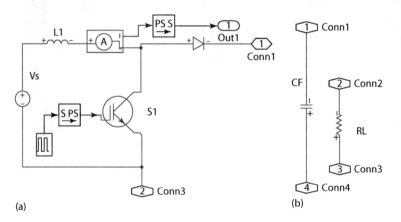

FIGURE 10.22
Boost converter model subsystems.

simulation results are tabulated in Table 10.8 for all values of the firing angle. Comparison of values in Table 10.8 with Tables 6.2 and 6.3 reveals that the error is small for α in the range from $\pi/6$ to $\pi/2$ rad and is large for $2\pi/3$ rad.

10.5.2 Three-Phase Thyristor AC to AC Controller in Series with Resistive Load in Delta

The component-level model of the three-phase thyristor AC to AC controller in series with resistive load in delta is shown in Figure 10.34 [2–5,8]. The interactive blocks of this model and its subsystems are shown in Figures 10.35 and 10.36, respectively. The model parameters are given in Table 6.4. The

Interactive Model Verification

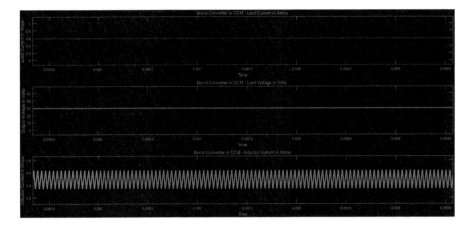

FIGURE 10.23
Boost converter in CCM: simulation results.

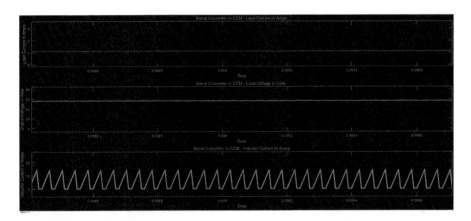

FIGURE 10.24
Boost converter in DCM: simulation results.

TABLE 10.6

Boost Converter: Simulation Results

Sl. No.	Output Voltage V_O (V)	Minimum Inductor Current (A)	Maximum Inductor Current (A)	Load Current (A)	Remarks
1	29.5	1.341	1.636	0.59	CCM
2	56.3	0	14.16	1.126	DCM

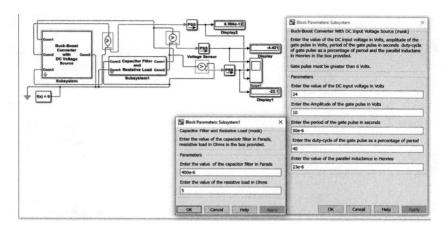

FIGURE 10.25
Buck–boost converter model.

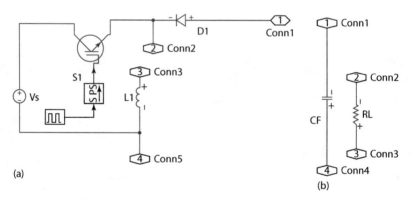

FIGURE 10.26
Buck–boost converter model subsystems.

simulation results are given in Figures 10.37 and 10.38, respectively. The simulation results are tabulated in Table 10.9. Comparison of the results in Table 10.9 with Tables 6.5 and 6.6 reveals an error in the circuit model. Compared with the calculated values in Table 6.6, the percentage error for $\pi/6$ rad is 9.26%, for $\pi/4$ rad it is 14.6% and so on. For π rad, the output voltage is 41 mV, whereas it should be zero. Here, non-ideal semiconductor components are used in the model that have forward voltage and on resistance.

10.6 Switched Mode Power Supply Using Buck Converter

The system model and simulation of a switched mode power supply (SMPS) using a buck converter with and without using a proportional–integral (PI) controller is presented in Chapter 7. Here, the component-level model of the

Interactive Model Verification

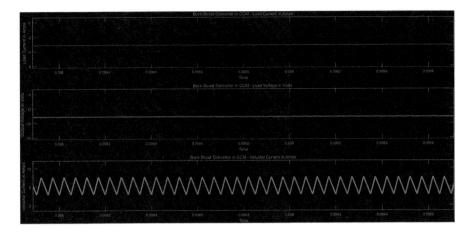

FIGURE 10.27
Buck–boost converter in CCM: simulation results.

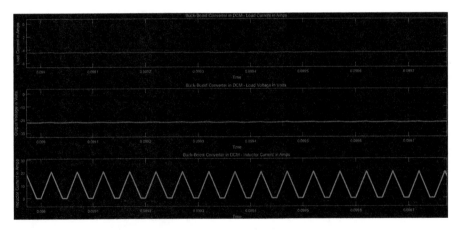

FIGURE 10.28
Buck–boost converter in DCM: simulation results.

TABLE 10.7

Buck–Boost Converter: Simulation Results

Sl. No.	Output Voltage V_O (V)	Minimum Inductor Current (A)	Maximum Inductor Current (A)	Load Current (A)	Duty Cycle	Remarks
1	−15.65	2.853	7.613	3.13	0.4	CCM
2	−22.13	0	20.69	4.426	0.4	DCM

274 Power Electronic Converters

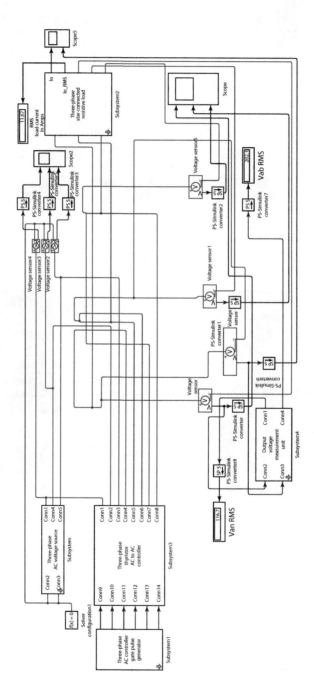

FIGURE 10.29
Three-phase thyristor AC to AC controller in series with star-connected resistive load.

Interactive Model Verification

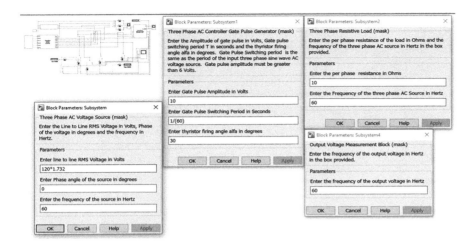

FIGURE 10.30
Three-phase thyristor AC to AC controller in series with star-connected resistive load with interactive blocks.

same SMPS is developed using semiconductor and passive components; a simulation was carried out and compared with that presented in Section 7.3 [6,7].

This model of the buck converter SMPS with and without PI controller is shown in Figure 10.39. The interactive blocks of this model are shown in Figure 10.40 and subsystems are shown in Figure 10.41.

The relevant data are entered in the box shown in the interactive blocks in Figure 10.40. In Figure 10.41, subsystems A, C and D form the buck converter, capacitor filter and load resistor and the output voltage and current measurement unit. Figure 10.41b is the gate pulse generating unit. The output voltage across the load resistor is multiplied by a potentiometer constant using a gain pot. The output of the gain pot is compared with a reference voltage v_{ref} in an error detector unit that gives the error output voltage V_{ERR}. The triangle carrier generator generates a triangle carrier with a positive and negative peak of +1 and −1 V and frequency corresponding to the switching frequency of the SMPS. The triangle carrier is multiplied by a gain multiplier constant gm to obtain the triangle carrier with required amplitude. The required amplitude of the triangle carrier is that which gives normal output voltage of 9 V DC with a normal input voltage of 15 V DC. The error output V_{ERR} is compared with the triangle carrier either directly or through the PI controller in a comparator and the resulting output of the comparator is the gate pulse for the SMPS. Switch SS1 selects V_{ERR} directly or through the PI controller.

The component model simulation of the buck SMPS is carried out using data given in Table 7.1. The model simulation results are shown in Figure 10.42a–c for the buck SMPS without the PI controller and in Figure 10.43a–c for the

276 Power Electronic Converters

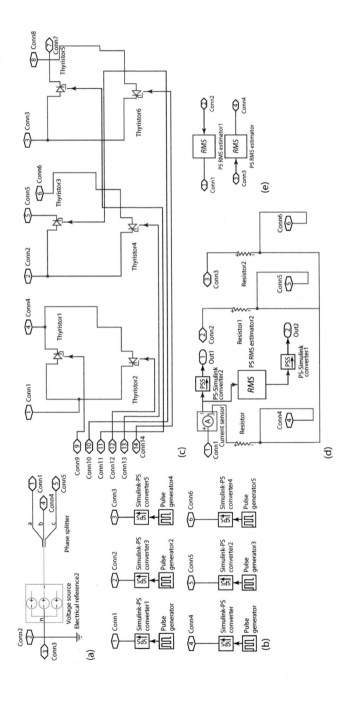

FIGURE 10.31
Three-phase thyristor AC to AC controller in series with star-connected resistive load model subsystems.

Interactive Model Verification

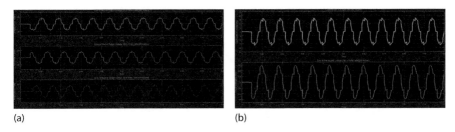

(a)   (b)

FIGURE 10.32
Three-phase thyristor AC to AC controller connected to (a) star-connected load and (b) star-connected resistive load: simulation results for $\alpha = \pi/6$ rad.

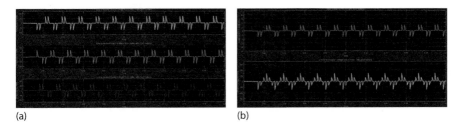

(a)   (b)

FIGURE 10.33
Three-phase thyristor AC to AC controller connected to (a) star-connected resistive load and (b) star-connected load: simulation results for $\alpha = 2\pi/3$ rad.

case with the PI controller. The simulation results for the buck SMPS without the PI controller and those for the case with the PI controller are tabulated in Table 10.10. The conclusions drawn from the results for the buck SMPS are given as follows:

- The change in output voltage for a given change in input voltage or regulation is slightly higher for the case with the PI controller than without the PI controller.

TABLE 10.8

Three-Phase Thyristor AC Controller Connected to Resistive Star-Connected Load: Simulation Results

Sl. No.	Line-to-Neutral Voltage (V)	Frequency (Hz)	Firing Angle α (rad)	Per Phase Load Resistance (Ω)	RMS Line-to-Neutral Load Voltage (V)	RMS Load Current (A)
1	120	60	$\pi/6$	10	116.2	11.62
2	120	60	$\pi/4$	10	110.8	11.08
3	120	60	$\pi/3$	10	100.8	10.08
4	120	60	$\pi/2$	10	64.39	6.439
5	120	60	$2\pi/3$	10	24.55	2.455

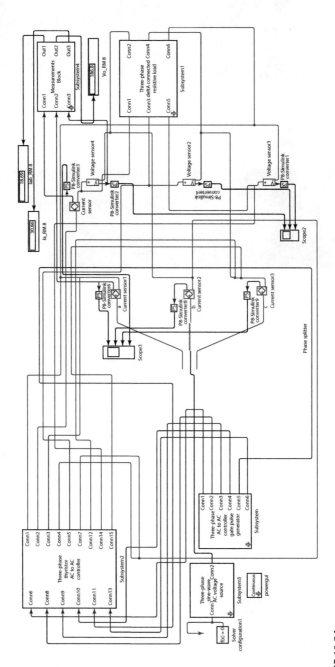

FIGURE 10.34
Three-phase thyristor AC to AC controller in series with resistive load in delta.

Interactive Model Verification 279

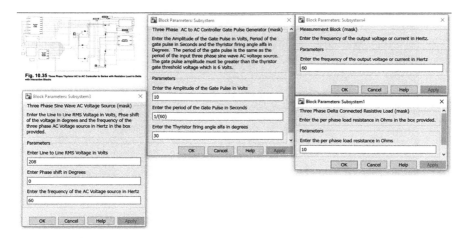

FIGURE 10.35
Three-phase thyristor AC to AC controller in series with resistive load in delta with interactive blocks.

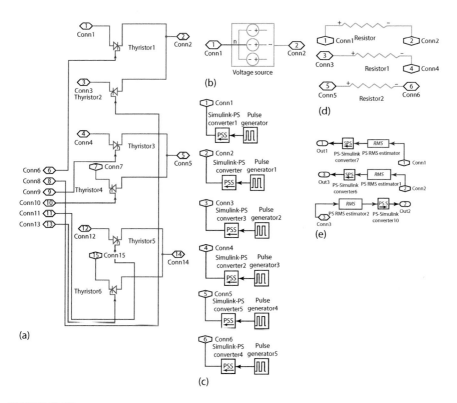

FIGURE 10.36
Three-phase thyristor AC controller in series with resistive load in delta: model subsystems.

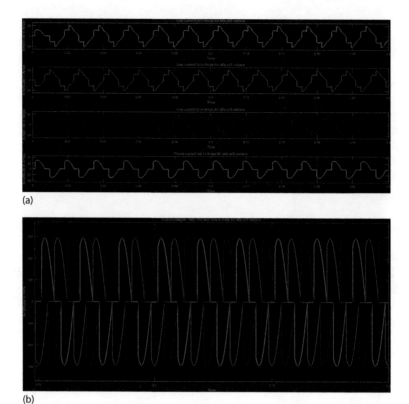

FIGURE 10.37
(a and b) Three-phase thyristor AC controller in series with resistive load in delta: simulation results for $\alpha = \pi/6$ rad.

- The output voltage is 9 V when the input voltage is 15 V in both cases.
- The number of initial oscillations for the first 0.01 s is nine without the PI controller and six with the PI controller.
- The initial peak overshoot was 16.25 V without the PI controller and 14.2 V with the PI controller.
- There is a reduction of peak overshoot by 2 V with the PI controller compared with the case without the PI controller.

10.7 Fourth-Order DC to DC Converters

The system modelling of selected fourth-order converters such as the single-ended primary inductance converter (SEPIC), quadratic boost and ultra-lift Luo converters are discussed in Chapter 8 for both CCM and DCM. In this

Interactive Model Verification 281

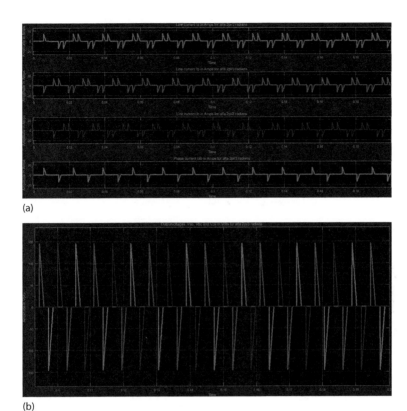

FIGURE 10.38
(a and b) Three-phase thyristor AC controller in series with resistive load in delta: simulation results for α = 2π/3 rad.

TABLE 10.9

Three-Phase AC Controller in Series with Resistive Load in Delta: Simulation Results

Sl. No.	RMS Line-to-Line Supply Voltage (V)	Frequency(Hz)	Firing α (rad)	Per Phase Load Resistance (Ω)	RMS Load Voltage V_O (V)	RMS Load Current I_{ab} (A)
1	208	60	π/6	10	186.0	18.6
2	208	60	π/4	10	168.8	16.88
3	208	60	π/3	10	146.6	14.66
4	208	60	π/2	10	91.55	9.1555
5	208	60	2π/3	10	35.03	3.503
6	208	60	π	10	0.041	0.0041

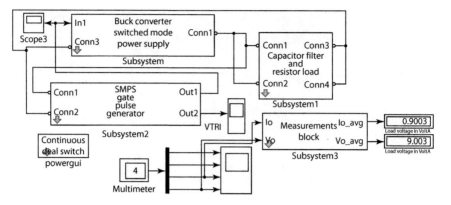

FIGURE 10.39
Model of buck converter SMPS.

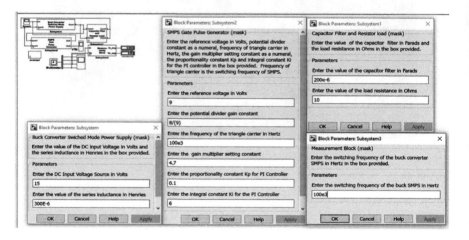

FIGURE 10.40
Model of buck converter SMPS with interactive blocks.

section, component-level models are developed for these three fourth-order converters using semiconductor and passive components, and simulation results for both CCM and DCM operation are presented, tabulated and compared with those given in Chapter 8 [6,7,9].

10.7.1 SEPIC Converter

The component-level model of the SEPIC converter is shown in Figure 10.44. The interactive blocks of this model and the subsystems are shown in Figures 10.45 and 10.46, respectively [6,7]. The model simulation results for CCM and DCM are shown in Figures 10.47 and 10.48, respectively.

Interactive Model Verification 283

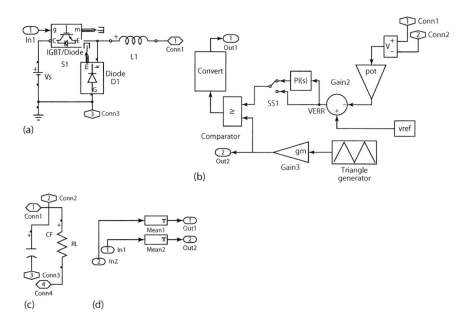

FIGURE 10.41
Buck converter SMPS model subsystems.

The data shown in Table 8.1 are used for simulation. The simulation results are also tabulated in Table 10.11. Comparison of the simulation results in Table 10.11 with those in Tables 8.2 and 8.3 reveals that the values for CCM almost agree. For the case of DCM, there are some discrepancies with the minimum value for iL_1 and maximum value for iL_2 when compared with those in Tables 8.2 and 8.3. The polarities of measurement of the iL_2 inductor current and V_{C1} voltage are reversed and, hence, a minus and plus sign are, respectively, inserted in Table 10.11.

10.7.2 Quadratic Boost Converter

The component-level model of the quadratic boost converter is shown in Figure 10.49 and model subsystems are shown in Figure 10.50 [6,7]. The simulation results are shown in Figure 10.51 for CCM and Figure 10.52 for DCM and tabulated in Table 10.12. Data shown in Table 8.4 are used for simulation. Comparison of simulation results in Table 10.12 with those in Table 8.5 shows that the values agree closely for CCM and for DCM, except for V_O. The V_O value has an error of 7.68%.

10.7.3 Ultra-Lift Luo Converter

The component-level model of the ultra-lift Luo converter with interactive blocks is shown in Figure 10.53 and the model subsystems are shown in

284   Power Electronic Converters

FIGURE 10.42
Buck converter SMPS without PI controller for a DC input voltage of (a) 15 V, (b) 12 V and (c) 18 V.

Interactive Model Verification

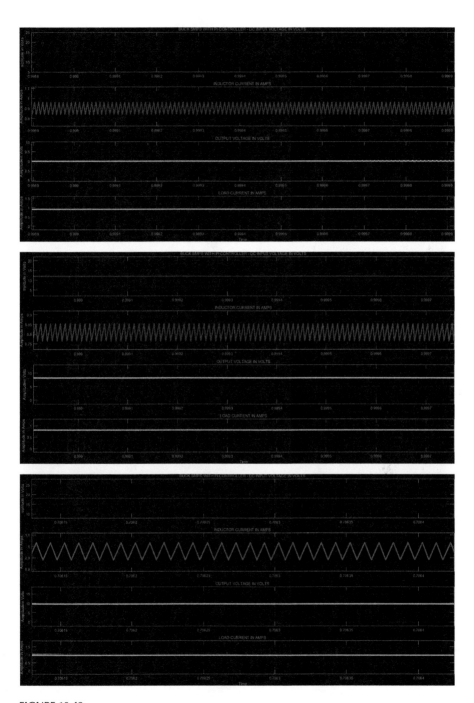

FIGURE 10.43
Buck converter SMPS with PI controller for a DC input voltage of (a) 15 V, (b) 12 V and (c) 18 V.

TABLE 10.10
Buck Converter SMPS Without PI Controller: Simulation Results

Sl. No.	PI Controller	DC Input Voltage (V)	DC Output Voltage (V)	Load Current (A)	Remarks
1	NO	15	9.007	0.9007	Normal input voltage
2	NO	12	8.163	0.8163	Lower input voltage
3	NO	18	9.674	0.9674	Higher input voltage
4	YES	15	9.003	0.9003	Normal input voltage
5	YES	12	8.08	0.808	Lower input voltage
6	YES	18	9.69	0.969	Higher input voltage

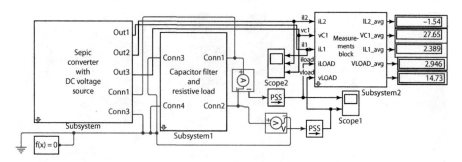

FIGURE 10.44
SEPIC converter model.

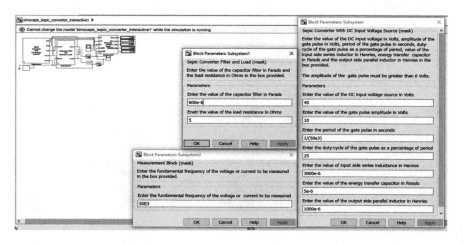

FIGURE 10.45
SEPIC converter model with interactive blocks.

Interactive Model Verification

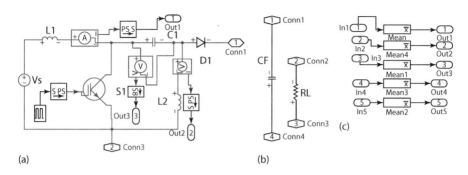

(a)   (b)

FIGURE 10.46
SEPIC converter model subsystems.

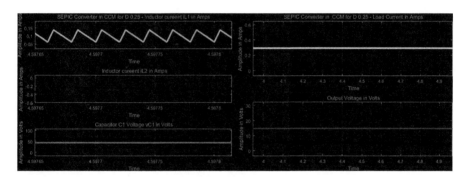

FIGURE 10.47
SEPIC converter in CCM for $D = 0.25$: simulation results.

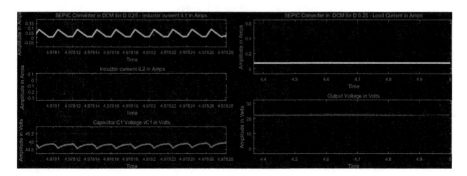

FIGURE 10.48
SEPIC converter in DCM for $D = 0.25$: simulation results.

TABLE 10.11

SEPIC Converter: Simulation Results

Sl. No.	Duty Ratio	Minimum L_1 Inductor Current (A)	Maximum L_1 Inductor Current (A)	Minimum L_2 Inductor Current (A)	Maximum L_2 Inductor Current (A)	Capacitor C_1 Voltage VC_1 (V)	Output Voltage V_O (V)	Load Current I_O (A)	Remarks
1	0.25	0.0624	0.137	−0.405	−0.181	44.99	14.73	0.294	CCM
2	0.25	0.0088	0.0875	−0.215	0.0132	44.92	22.19	0.0739	DCM

Interactive Model Verification 289

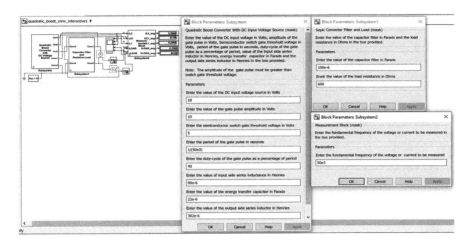

FIGURE 10.49
Quadratic boost converter model with interactive blocks.

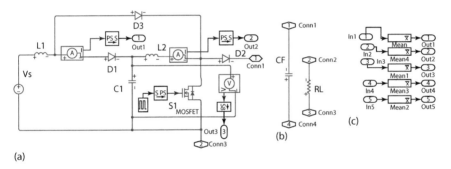

FIGURE 10.50
Quadratic boost converter model subsystems.

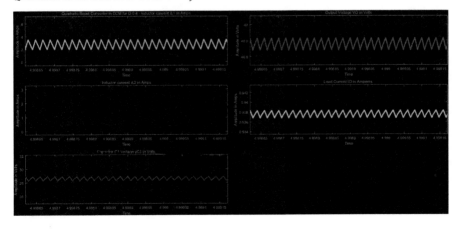

FIGURE 10.51
Quadratic boost converter in CCM for $D = 0.4$: simulation results.

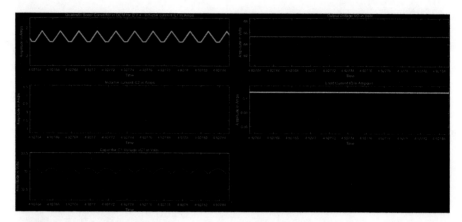

FIGURE 10.52
Quadratic boost converter in DCM for $D = 0.4$: simulation results.

Figure 10.54 [6,7]. Simulation of this model was carried out for both CCM and DCM using the data shown in Table 8.7. The simulation results for CCM and DCM are shown in Figures 10.55 and 10.56, respectively.

The simulation results are tabulated in Table 10.13. Comparison of the simulation results in Table 10.13 with those in Tables 8.8 and 8.9 reveals that the maximum and minimum values of the two inductor currents iL_1 and iL_2 and the load current I_O agree closely. Capacitor voltage V_{C1} for CCM and DCM and output voltage V_O for CCM have some discrepancies. The output voltage V_O for DCM agrees closely with the calculated value in Table 8.9. The current measurement direction for iL_2 and I_O is reversed and, hence, a minus sign is inserted in Table 10.13.

10.8 Three-Phase Three-Level Inverters

The system modelling of the three-phase diode-clamped three-level inverter (DCTLI) and the flying-capacitor three-level inverter (FCTLI) are reported in Sections 9.2 and 9.3. These system models are developed using ideal switches and Pulse Generator blocks, which signify the behaviour of three-phase DCTLI and FCTLI. In this chapter, component-level models are developed for three-phase DCTLI and FCTLI, simulation results obtained and compared with those presented in Chapter 9 [10–13]. These models are developed using passive and semiconductor components, which are not ideal switches; that is, these semiconductor components have forward voltage drop and on and off resistance.

TABLE 10.12

Quadratic Boost Converter: Simulation Results

Sl. No.	Duty Ratio	Minimum L_1 Inductor Current (A)	Maximum L_1 Inductor Current (A)	Minimum L_2 Inductor Current (A)	Maximum L_2 Inductor Current (A)	Capacitor C_1 Voltage VC_1 (V)	Output Voltage V_O (V)	Load Current I_O (A)	Remarks
1	0.4	1.985	3.51	1.296	1.894	28.6	46.88	0.9375	CCM
2	0.4	0	1.53	0	0.6915	33.05	65.31	0.1088	DCM

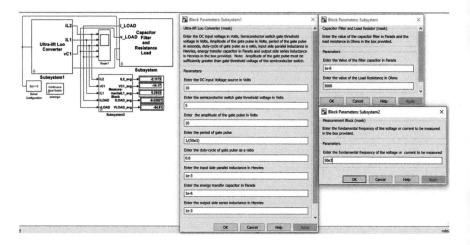

FIGURE 10.53
Ultra-lift Luo converter model with interactive blocks.

10.8.1 Three-Phase Diode-Clamped Three-Level Inverter

Detailed analysis of the three-phase DCTLI is shown in Section 9.2. The circuit topology is given in Figure 9.1 [10–13]. The component-level model of the three-phase DCTLI is shown in Figure 10.57.

The interactive blocks of this model are shown in Figure 10.58 and the model subsystems are shown in Figure 10.59.

Figure 10.59a–e shows, respectively, semiconductor switches with clamping diodes, the gate pulse generator unit, the DC voltage source with DC link capacitors, three-phase load resistors and the output voltage root mean square (RMS) value and total harmonic distortion (THD) measurement unit.

Simulation results are shown in Figure 10.60.

Simulation results are tabulated in Table 10.14. Comparison of the simulation results in Table 10.14 with the results and derivation in Chapter 9 shows close agreement.

10.8.2 Three-Phase Flying-Capacitor Three-Level Inverter

The three-phase FCTLI system model is discussed in Section 9.3. Figure 9.8 shows the circuit topology of the three-phase FCTLI [10–13]. The component-level model using passive and semiconductor components is shown in Figure 10.61.

The interactive blocks of this model are shown in Figure 10.62 and the model subsystems are shown in Figure 10.63.

Interactive Model Verification

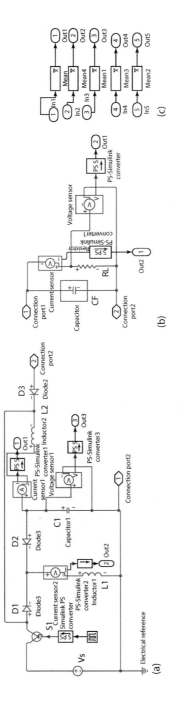

FIGURE 10.54
Ultra-lift Luo converter model subsystems.

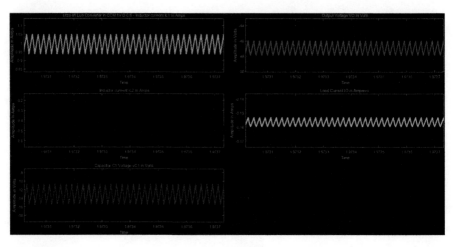

FIGURE 10.55
Ultra-lift Luo converter in CCM: simulation results.

Figure 10.63a–e shows, respectively, the three-phase semiconductor switches with flying capacitors, the gate pulse generators, the DC voltage source, the three-phase resistor load and the output voltage RMS and THD measurement unit.

A small resistance of 1 µΩ is connected in series with each DC voltage source for the convergence of simulation. The simulation results are shown in Figure 10.64.

Simulation results are tabulated in Table 10.15. Comparison of the simulation results for the three-phase FCTLI with those in Chapter 9 and also with the derived formula shows that the results in Table 10.15 agree closely.

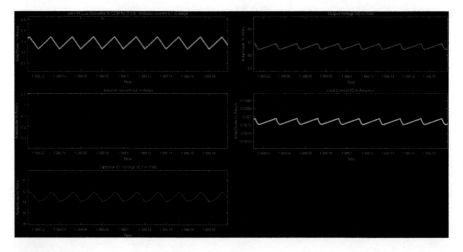

FIGURE 10.56
Ultra-lift Luo converter in DCM: simulation results.

TABLE 10.13

Ultra-Lift Luo Converter: Simulation Results

Sl. No.	Duty Ratio	Minimum L_1 Inductor Current (A)	Maximum L_1 Inductor Current (A)	Minimum L_2 Inductor Current (A)	Maximum L_2 Inductor Current (A)	Capacitor C_1 Voltage VC_1 (V)	Output Voltage V_O (V)	Load Current I_O (A)	Remarks
1	0.6	0.9378	1.048	−0.256	−0.536	−13.19	−47.02	−0.156	CCM
2	0.6	0.2291	0.3395	−0.2795	0	−13.09	−81.42	−0.027	DCM

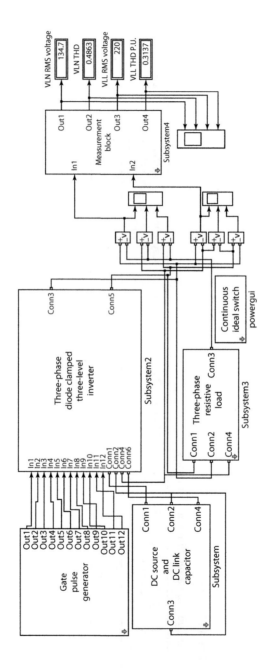

FIGURE 10.57
Three-phase DCTLI model.

Interactive Model Verification

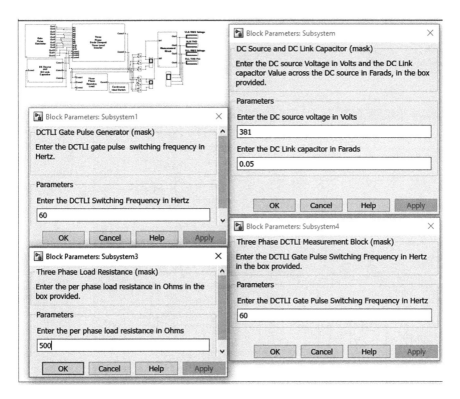

FIGURE 10.58
Three-phase DCTLI model with interactive blocks.

10.9 Three-Phase Sine PWM Inverter

The method of generating gate pulses for a three-phase inverter by the sine PWM technique is discussed in Section 4.4. The system model of the three-phase sine PWM inverter is presented in Section 4.4.1. In this section, the circuit component-level model of the three-phase sine PWM inverter is developed and the results obtained by simulation are compared with those presented in Chapter 4.

The interactive component-level model of the three-phase sine PWM inverter is shown in Figure 10.65.

The model subsystems are shown in Figure 10.66a–d.

In Figure 10.65, the data entered are for an overmodulation index of 1.1. The DC link voltage is 286 V, the sine-wave modulating signal frequency is 60 Hz, the triangle carrier frequency is 2 kHz and the load resistance is 1 kΩ. The MOSFET gate threshold voltage is 5 V and, therefore, the gain multiplier for the gate pulse amplitude entered is 10, so as to exceed this gate threshold voltage. Two amplitude modulation (A.M.) indices are used for simulation,

298 Power Electronic Converters

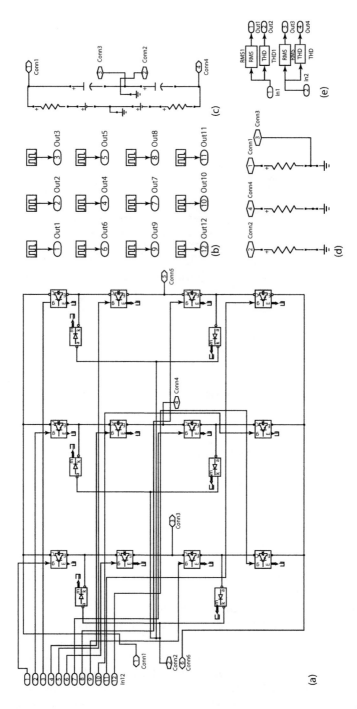

FIGURE 10.59
Three-phase DCTLI model subsystems.

Interactive Model Verification

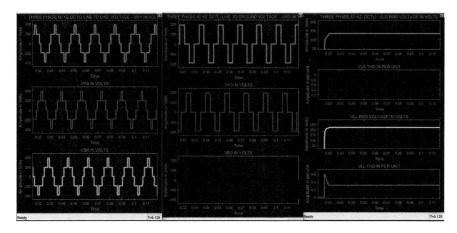

FIGURE 10.60
Three-phase DCTLI: simulation results.

TABLE 10.14

Three-Phase DCTLI: Simulation Results

Sl. No.	DC Link Voltage (V)	Frequency (Hz)	V_{LL} RMS (V)	V_{LL} THD Per Unit	V_{LG} RMS (V)	V_{LG} THD Per Unit
1	381	60	220	0.3137	134.7	0.4863

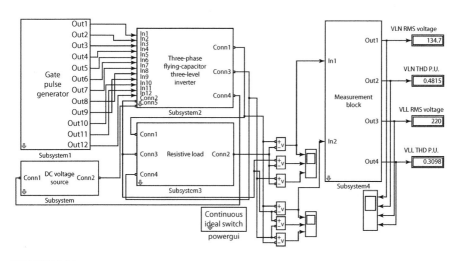

FIGURE 10.61
Three-phase FCTLI model.

FIGURE 10.62
Three-phase FCTLI model with interactive blocks.

which is the ratio of the peak of the modulating sine wave to the peak of the triangle carrier wave. The modulating sine-wave amplitude is entered as 1.1 V. The modulating sine wave has an amplitude between 0 and 1 for undermodulation, 1 for normal modulation and above 1 for overmodulation.

Figure 10.66a shows the three-phase sine-triangle carrier PWM generator and Figure 10.66b shows the three-phase MOSFET inverter model subsystems. The three-phase sine-wave modulating signals V_{an}, V_{bn} and V_{cn} are compared separately with a triangle carrier having a peak value of 1 V using Relational Operator blocks known as *comparators*, the output of which form the PWM gate pulse for the three upper arm switches S1, S3 and S5, respectively. These three PWM gate pulses are then inverted using NOT gates to generate the PWM gate pulse for the lower arm switches S2, S4 and S6, respectively. Figure 10.66c shows the load current, line-to-neutral and line-to-line output voltage measurement subsystem. The RMS value and THD measurement subsystem for the line-to-neutral and line-to-line output voltages are shown in Figure 10.66d.

The component-level model simulation of the three-phase sine PWM inverter was carried out in Simulink using the ode23t (mod. stiff/trapezoidal) solver for two values of AM indices, 0.9 and 1.1. The simulation results are shown in Figure 10.67a–h for the former and Figure 10.68a–h for the latter AM index.

The results are tabulated in Table 10.16.

10.10 Three-Phase Five-Level Cascaded H-Bridge Inverter

The topology and principle of operation of the three-phase five-level cascaded H-bridge inverter (TPFLCHBI) is presented in Section 9.4 and the interactive system model development is presented in Section 9.4.1. Here, the interactive

Interactive Model Verification

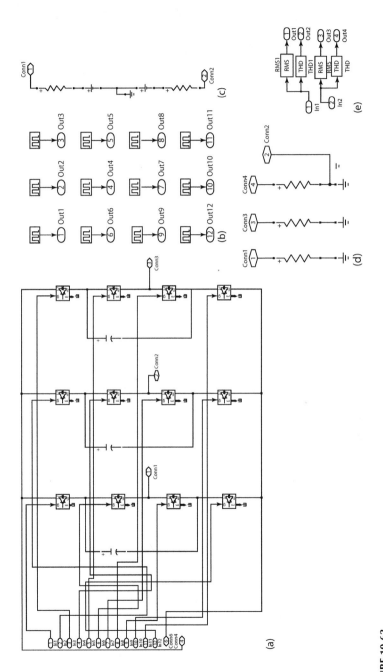

FIGURE 10.63
Three-phase FCTLI model subsystems.

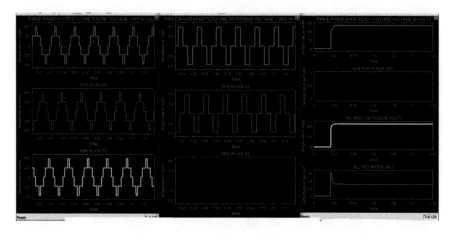

FIGURE 10.64
Three-phase FCTLI: simulation results.

TABLE 10.15

Three-Phase FCTLI: Simulation Results

Sl. No.	DC Link Voltage (V)	Frequency(Hz)	RMS (V)	THD Per Unit	RMS (V)	THD Per Unit
1	381	60	220	0.3098	134.7	0.4815

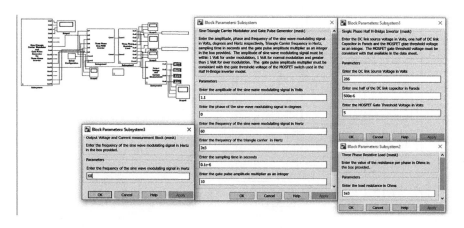

FIGURE 10.65
Three-phase sine PWM inverter with interactive blocks.

Interactive Model Verification

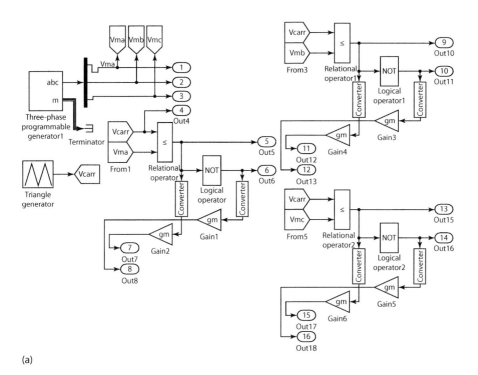

(a)

FIGURE 10.66
Three-phase sine PWM inverter model subsystems. (a) Three-phase sine-triangle carrier PWM generator.

component-level model of this converter is presented. Model simulation results were obtained and are presented and compared with those given in Chapter 9.

The component-level model of the TPFLCHBI is shown in Figure 10.69 [14,15]. This model, with an interactive dialogue box, is shown in Figure 10.70.

The internal subsystems of this model are shown in Figure 10.71a–d. In this model, MOSFET semiconductor switches are used. In the dialogue box in Figure 10.70, the upper and lower H-bridge DC voltage source, amplitude of the gate pulse, gate pulse switching period, load resistance, load inductance, switching frequency and MOSFET gate threshold voltage are entered in the appropriate box. In Figure 10.71a–d, the voltage source for the upper and lower H-bridge, gate pulse generator for the MOSFET switches, MOSFET switch arrangement for the TPFLCHBI and the R–L load with measurement blocks are shown.

Here, the MOSFET gate threshold voltage is entered as 5 V and, hence, a gate pulse amplitude of 10 V is used. The pulse generator shown in Figure 10.71b is the same as that shown in Figure 9.18a. Here, the pulse amplitude A is 10 V and the period T is $1/(50)$ s. The method of calculating the duty cycle and phase delay for various switches is the same as explained in Section 9.4.1 and in Table 9.8. The R–L load component values are 50 Ω and 0.5 H.

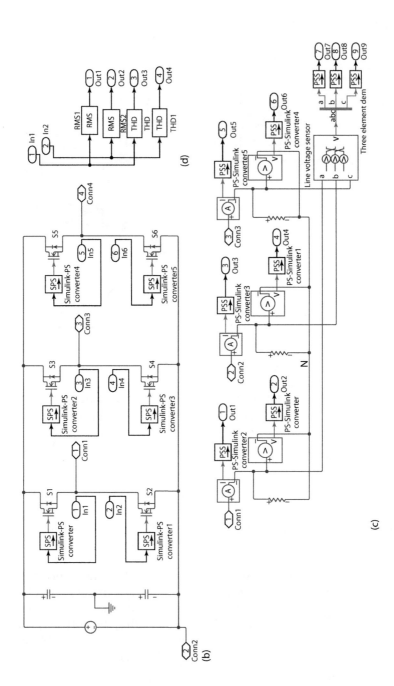

FIGURE 10.66
Three-phase sine PWM inverter model subsystems. (b) Three-phase MOSFET inverter. (c) Output voltage and current measurement unit. (d) Resistive load.

Interactive Model Verification

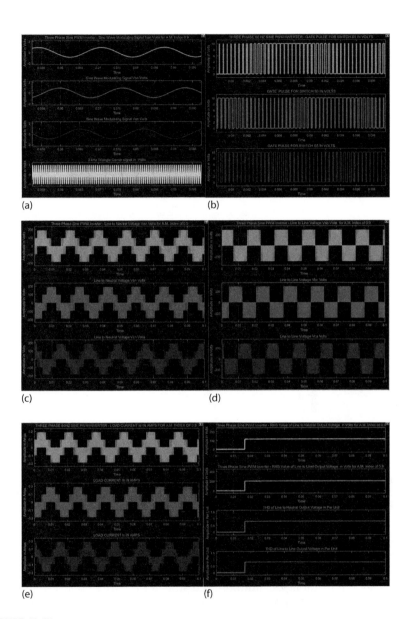

FIGURE 10.67
Three-phase sine PWM inverter component-level simulation. (a) Three-phase sine-wave modulating signal and triangle carrier (left). (b) Three-phase PWM gate pulse for upper arm switches of inverter (right). (c) Three-phase line-to-neutral output voltage (left). (d) Three-phase line-to-line output voltage (right). (e) Three-phase load current (left). (f) RMS value and THD of line-to-neutral and line-to line output voltage (right).

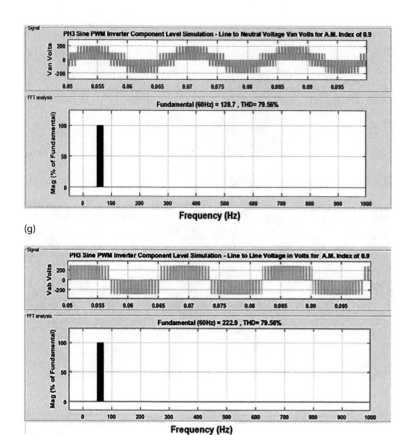

FIGURE 10.67 (CONTINUED)
(g) Harmonic spectrum of line-to-neutral voltage and (h) harmonic spectrum of line-to-line voltage.

The simulation for the component-level model of TPFLCHBI was carried out using the ode23t (mod. Stiff/Trapezoidal) solver [1]. The data used are the same as shown in Table 9.9. Simulation results for the three phase-to-ground voltages, load currents, RMS value and THD of phase-to-ground voltage are shown in Figure 10.72.

The harmonic spectrum of the phase-to-ground voltage is shown in Figure 10.73.

Simulation results are tabulated in Table 10.17.

Simulation results in Table 10.17 agree well with those shown in Table 9.9 and the derivation in Section 9.6.

Interactive Model Verification

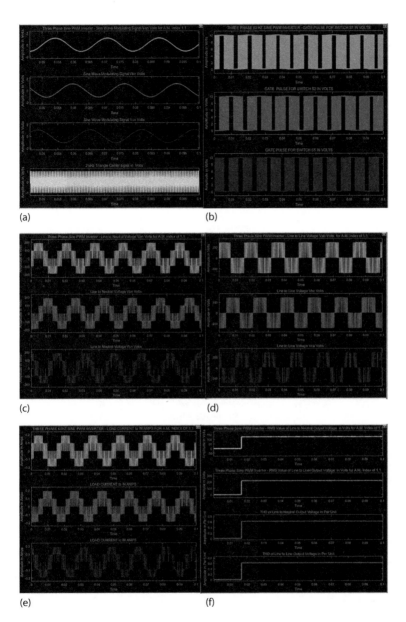

FIGURE 10.68
(a–f) Three-phase sine PWM inverter component level simulation. (a) Three-phase sine modulating signal and triangle carrier (left). (b) Three-phase PWM gate pulse for upper arm switches of inverter (right). (c) Three-phase line-to-neutral output voltage (left). (d) Three-phase line-to-line output voltage (right). (e) Three-phase load current (left). (f) RMS value and THD of line-to-neutral and line-to-line output voltages (right).

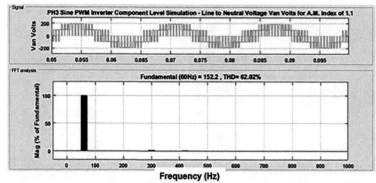

(g)

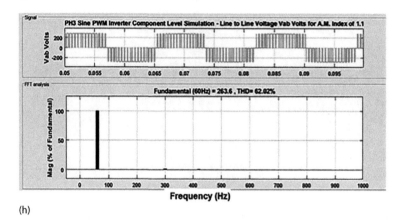

(h)

FIGURE 10.68 (CONTINUED)
(g, h) Three-phase sine PWM VSI component-level simulation: harmonic spectrum of (g) V_{an} and (h) V_{ab} for AM index of 1.1.

TABLE 10.16

Three-Phase Sine PWM Inverter: Component-Level Simulation Results

Sl. No.	DC Link Voltage (V)	AM Index	RMS V_{ab} (V)	THD V_{ab} Per Unit	RMS V_{an} (V)	THD V_{an} Per Unit
1	286	0.9	201.8	0.7923	116.3	0.7944
2	286	1.1	219.8	0.6178	126.6	0.6187

Interactive Model Verification

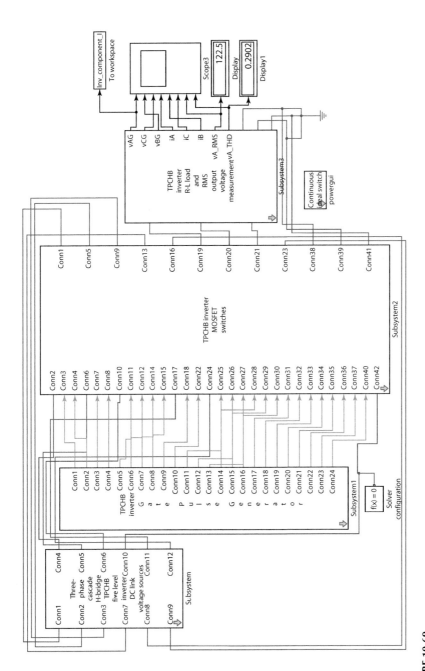

FIGURE 10.69
TPFLCHBI model.

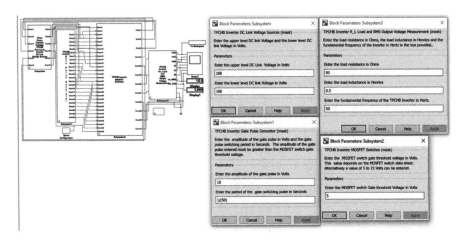

FIGURE 10.70
TPFLCHBI model with interactive blocks.

10.11 Pulse Width Modulation Methods for Multilevel Converters

Multi-carrier-based sine PWM techniques are classified as multi-carrier sine phase-shift PWM (MSPSPWM) and multi-carrier sine level shift PWM (MSLSPWM). These two PWM techniques for multilevel converters have already been presented in Section 9.7. In this section, details regarding the interactive component-level model development for these two PWM techniques are presented with model simulation results.

10.11.1 Multi-Carrier Sine Phase-Shift PWM

The system model of MSPSPWM is explained in Section 9.7.1. Here, the component-level model of MSPSPWM is presented. Figure 10.74 presents the component-level model of the MSPSPWM with an interactive dialogue box.

Figure 10.75 shows the internal subsystem with a dialogue box. In the dialogue box in Figure 10.74, the numerical data relating to triangle carrier frequency, angular phase-shift between four carriers, modulating signal amplitude and frequency in hertz and phase difference between three-phase sine-wave modulating signal are entered. The triangle carrier has a peak value of +1 V/–1 V.

In the model subsystem in Figure 10.75, for 'Triangle Generator' to 'Triangle Generator4', the frequency 'ftri', and phases 'phi*0' to 'phi*4' are entered. In the sine-wave generators marked 'Sine Wave1' to 'Sine Wave3', the amplitude is entered as 'A', the frequency is '2*pi*fm' rad/s and the phases 'theta*0*pi/180', 'theta*(–1)*pi/180' and 'theta*(–2)*pi/180' rad are entered. Phase shift 'phi' 90° creates four triangle carriers with a phase difference of 90° and peak

Interactive Model Verification

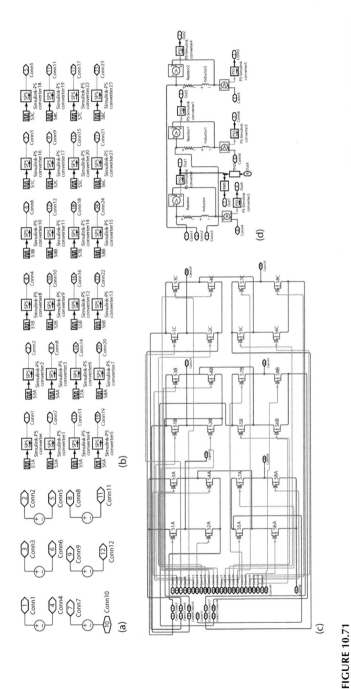

FIGURE 10.71
(a–d) TPFLCHBI model subsystems. (a) Voltage sources. (b) TPFLCHBI gate pulse generator. (c) TPFLCHBI MOSFET switches. (d) TPFLCHBI R–L load.

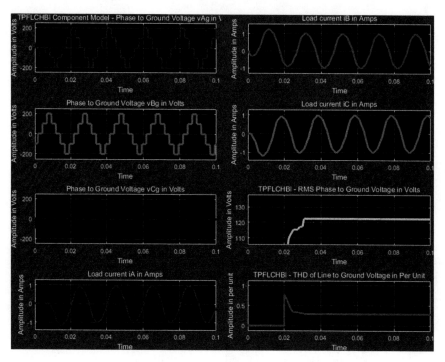

FIGURE 10.72
TPFLCHBI component-level model: simulation results.

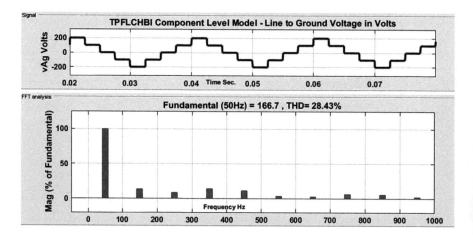

FIGURE 10.73
TPFLCHBI component-level model: harmonic spectrum of line-to-ground voltage.

Interactive Model Verification

TABLE 10.17

TPFLCHBI: Component-Level Simulation Results

Sl. No.	Symmetrical DC Link Voltage Source (V)	Frequency (Hz)	V_{Ag} RMS (V)	V_{Ag} THD (%)
1	100	50	122.5	29.02

value +1 V/−1 V. The ftri value is 10 kHz, which is the triangle carrier frequency in hertz. For the sine-wave modulating signal, amplitude 'A' corresponds to 0.8 V, frequency 'fm' corresponds to 50 Hz, and 'theta' corresponds to 120°.

The output of each triangle generator is compared with the three-phase sine-wave modulating signal using Relational Operator blocks, as shown in Figure 10.75. As the op-amp model is *not* functioning, Relational Operator

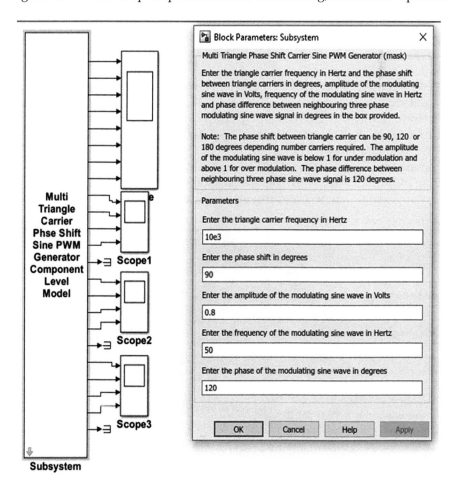

FIGURE 10.74
Component-level model of MSPSPWM with interactive block.

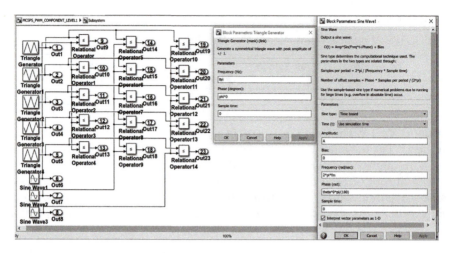

FIGURE 10.75
Subsystem of MSPSPWM component-level model.

blocks are used. Strictly, in a component-level model, the triangular carrier output and the sine-wave modulating signal output are passed to the inverting and non-inverting pin of an Op-Amp Comparator block. The resulting output of the Relational Operator/Op-Amp comparator is the PWM gate pulse for the multilevel converter.

Simulation of the MSPSPWM component-level model was carried out using the ode23t (mod. stiff/trapezoidal) solver [1]. The simulation results for the four triangle carriers and three-phase modulating signals are shown in

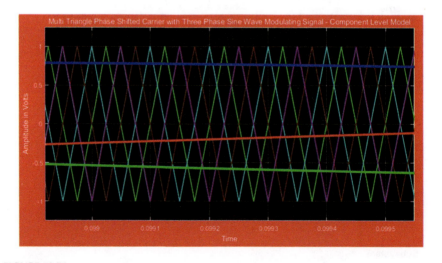

FIGURE 10.76
MSPSPWM component model: four triangle carriers and three-phase sine-wave modulating signal.

Interactive Model Verification 315

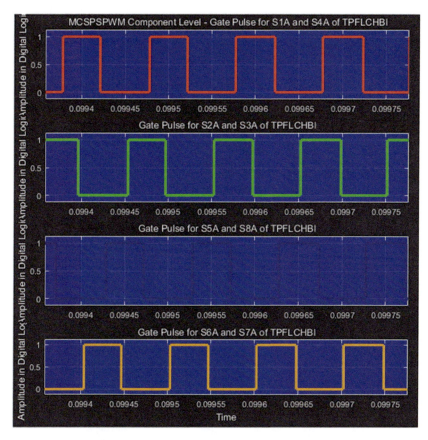

FIGURE 10.77
MSPSPWM component model: gate pulses for Phase A of TPFLCHBI.

Figure 10.76 and the gate pulses for Phase A of the TPFLCHBI are shown in Figure 10.77.

10.11.2 Multi-Carrier Sine Level Shift PWM

The details regarding the operation of the MSLSPWM method are explained in Section 9.7.3. In this section, details regarding the interactive component-level model of MSLSPWM are presented.

The model of the MSLSPWM generator is shown in Figure 10.78a–d with a dialogue box. Each model generates four different triangle carrier patterns. The internal subsystem of the model is shown in Figure 10.79 with a dialogue box, which corresponds to Figure 10.78b. In the dialogue box in Figure 10.78, the values entered are 10 kHz for the triangle carrier frequency, 1.5 V and 50 Hz for the sine-wave modulating signal amplitude and frequency, and 120° for the phase difference between three-phase sine-wave modulating signals.

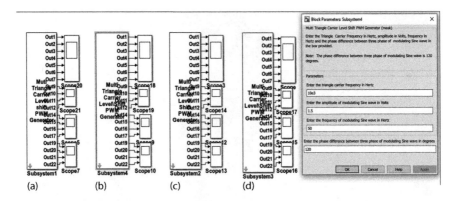

FIGURE 10.78
Multi-carrier level-shift PWM generator with interactive block.

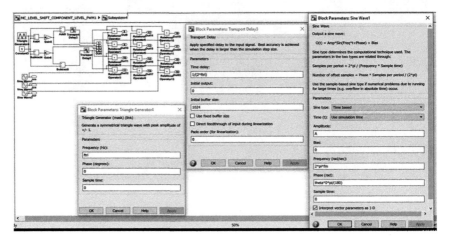

FIGURE 10.79
MSLSPWM generator subsystem.

Similarly, in Figure 10.79, in the dialogue box for 'Triangle Generator4', the frequency in hertz is entered as 'ftri'. The time delay in the 'Transport Delay2' and 'Transport Delay3' dialogue boxes is entered as '1/(2*ftri)'. The amplitude, frequency and phase are entered in the 'Sine Wave1' generator as 'A', '2*pi*fm' and 'theta*0*pi/(180)'. For Sine Wave2 and 3, the phases entered are 'theta*(–1)*pi/(180)' and 'theta*(–2)*pi/(180)' with amplitude and frequency values the same as for Sine Wave1. When 'ftri' is entered as 10 kHz, the triangle generator generates a triangle carrier with frequency of 10 kHz and amplitude of +1 V/–1 V. Transport Delay2 and 3 provide a time delay of '1/(2*10e3)' s. Sine Wave1, 2 and 3 output a three-phase sine wave with amplitude 1.5 V, frequency 'fm' corresponding to 50 Hz and phase 'theta' corresponding to zero, –120° and –240°, respectively. To obtain a triangle carrier with +1 V peak value

Interactive Model Verification

and zero minimum value, the Triangle Generator4 output is added to 1 with a constant block, and this output is multiplied by 0.5 using the Gain block. To obtain a triangle carrier with −1 V peak and zero maximum value, the Triangle Generator4 output is subtracted from 1 in a constant block and multiplied by 0.5 using the Gain block. The triangle carrier with +1 V peak and zero minimum value is again added to 1 from a constant block and delayed in time by '1/(2*ftri)' to obtain the next higher level triangle carrier. Similarly, the triangle carrier with −1 V peak and maximum value zero is subtracted from 1 in a constant block and delayed in time by '1/(2*ftri)' to obtain the next lower level triangle carrier. This generates four triangle carriers with Pattern B in Figure 10.78. To obtain Pattern A for the triangle carrier, the two transport delays are removed. To obtain Pattern C for the triangle carrier, Transport Delay2 is connected to the output of the Gain6 block and for Pattern D, Transport Delay3 is connected to the output of the Gain6 block in Figure 10.79.

The four triangle carrier outputs are compared with modulating sine waves in each phase using Relational Operator blocks. As the op-amp module

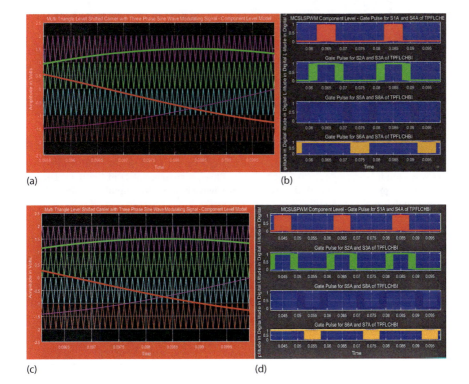

FIGURE 10.80
MSLSPWM component level model. (a) Four triangle carriers in Pattern A with three-phase sine-wave modulating signal and (b) gate pulse for Phase A switches of TPFLCHBI. (c) Four Triangle carrier in Pattern B with three-phase wave modulating saignal. (d) Gate pulse for Phase A of TPFLCHBI.

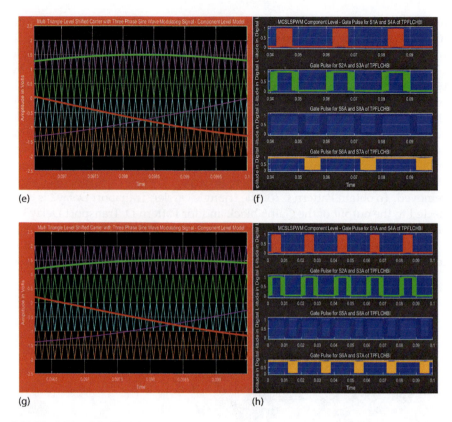

FIGURE 10.80 (CONTINUED)
MSLSPWM component level model. (e) Four triangle carrier in Pattern C with three-phase since wave mudulating signal. (d) Gate pulse for Phase A switches of TPFLCHBI. (g) Four triangle carriers in Patteren D with three phase since wave modulating signal. (h) Gate pulse for Phase A switches of TPFLCHBI.

is *not* functioning, Relational Operator blocks are used in this component model. The output of the Relational Operator blocks are the gate pulses to the switches in each phase of the multilevel converter.

The simulation of the MSLSPWM generator component-level model was carried out using the ode23t (mod. stiff/trapezoidal) solver [1]. The simulation results for Patterns A to D in Figure 10.78 are shown in Figure 10.80a–h.

10.12 Conclusions

This chapter is exclusively dedicated to developing interactive component-level models for the power electronic converters such as AC to DC, DC to AC, DC to DC, AC to AC, buck SMPS, fourth-order DC to DC converters

Interactive Model Verification 319

and three-phase DCTLI, FCTLI, TPFLCHBI, sine PWM, MSPSPWM and MSLSPWM, whose system models are presented and discussed in Chapters 3 to 9. The system models use ideal switches, whereas component-level models use semiconductor switches and components that are non-ideal. The results obtained by simulation of the component-level model for each power electronic converter are compared with those of their respective system model. The possible causes of discrepancies in the simulation results between these two types of model are presented.

References

1. The Mathworks Inc.: "MATLAB/Simulink Release Notes", R2016b, 2016.
2. M.H. Rashid: *Power Electronics Circuits, Devices and Applications*, Upper Saddle River, NJ: Pearson Education, Pearson Prentice Hall, 2004.
3. I. Batarseh: *Power Electronic Circuits*, Hoboken, NJ: Wiley, 2004.
4. D.W. Hart: *Introduction to Power Electronics*, Upper Saddle River, NJ: Prentice Hall, 1997.
5. N. Mohan, T.M. Undeland, and W.P. Robbins: *Power Electronics: Converters, Applications and Design*, Hoboken, NJ: Wiley, 1995.
6. B. Choi: *Pulse Width Modulated DC to DC Power Conversion: Circuit, Dynamics and Control Designs*, Piscataway, NJ: IEEE Press – Wiley, 2013, pp. 93–143.
7. F.L. Luo and H. Ye: *Power Electronics*, Boca Raton, FL: CRC Press, 2013.
8. M.H. Rashid [Ed.]: *Power Electronics Handbook*, Burlington, MA: Elsevier, 2011.
9. F.L. Luo and H. Ye: "Ultra-lift Luo converter", *IEE Proceedings on Electric Power Applications*; Vol.152, 2005; pp. 27–32.
10. P.M. Bhagwat and V.R. Stefanovic: "Generalized structure of a multilevel PWM inverter"; *IEEE Transactions on Industry Applications*; Vol.IA-19, No.6, November/December 1983; pp. 1057–1069.
11. A. Nabae, I. Takahashi, and H. Akagi: "A neutral-point clamped PWM inverter", *IEEE Transactions on Industry Applications*; Vol.IA-17, No.5, September/October 1981; pp. 518–523.
12. J.S. Lai and F.Z. Peng: "Multilevel convertors: A new breed of power convertors"; *IEEE Transactions on Industry Applications*; Vol. 32, No.3, May/June 1996; pp. 509–517.
13. J. Rodriguez, J.S. Lai and F.Z. Peng: "Multilevel inverters: A survey of topologies control and applications", *IEEE Transactions on Industrial Electronics*; Vol.49, No. 4, August 2002; pp. 724–738.
14. J.W. Dixon: "Multilevel converters" in *Power Electronic Converters and Systems Frontiers and Applications*, Ed. A.M. Trzynadlowski, London: The Institution of Engineering and Technology, 2016; pp. 43–72.
15. S. Khomfoi and L.M. Tolbert: "Multilevel power converters" in *Power Electronics Handbook*, Ed. M.H. Rashid, Burlington, MA: Elsevier, Butterworth-Heinemann, 2011; pp. 455–484.

11

Interactive Model for and Real-Time Simulation of a Single-Phase Half H-Bridge Sine PWM Inverter

11.1 Introduction

This chapter describes how Texas Instruments (TI) digital signal processor (DSP) TMS320F2812 can be used to generate pulse width modulation (PWM) gate drive output for the single-phase half H-bridge (SPHHB) inverter. This DSP has four main units: (1) internal and external bus; (2) central processing unit (CPU); (3) memory; and (4) peripherals [1]. The control algorithm is developed in MATLAB®/Simulink® using the Embedded coder block set; code generated using Real-Time Workshop (RTW) is compiled, debugged and transferred to the DSP board using Code Composer Studio (CCS) software.

CCS is an integrated design environment (IDE) that provides an environment for project development with all tools for real-time applications using only TI DSP. The program can be written either in Assembler or in C.

Experimental investigation of the gate drive, output voltage and current waveforms for an SPHHB inverter were studied using a Simulink model of the gate drive developed using Simulink and the Embedded coder block set interfaced to TI DSP TMS320F2812. Real-time simulation, known as a *software in the loop (SIL) simulation*, was conducted and the results are presented.

The experimental set-up consists of an eZdsp F2812 board, an SPHHB inverter and its gate drive modules. Here, the method of generating the PWM gate drive using a sine-triangle carrier is explained. This inverter was developed using silicon carbide (SiC) power MOSFETs.

321

11.2 Interactive Model of Single-Phase Half H-Bridge Sine PWM Inverter

The model of the SPHHB sine PWM inverter with interactive blocks developed using Simulink software is shown in Figure 11.1 [1–5]. These model subsystems are shown in Figure 11.2. Figure 11.2a–d are, respectively, sine-triangle carrier pulse width modulator, SPHHB MOSFET inverter, resistor and coupled coil inductor load, and output voltage and current measurement unit. In Figure 11.2a, the reference phase of the sine-wave modulating signal with frequency 400 Hz and peak value 0.5 V and a triangle carrier with frequency 20 kHz are compared in a relational operator block to generate gate pulses for the top switch S1 in Figure 11.2b. This output of the relational operator block is inverted using a NOT gate to generate gate pulses for the bottom switch S2 in Figure 11.2b. The DC link voltage is 60 V. The resistor, with value 15 Ω, and two coupled coil mutual inductors, each with 1 mH self-inductance and 0.9 coupling coefficient, shown in Figure 11.2c, form the load in series. Figure 11.2d shows the root mean square (RMS) output voltage and current measurement unit.

11.2.1 Simulation Results

Simulation of the model was carried out using the ode23t (mod. stiff/trapezoidal) solver [5]. The model simulation results are shown in Figure 11.3a–e, respectively.

11.3 Real-Time Software in the Loop Simulation

The real-time SIL simulation of the SPHHB sine PWM inverter was carried out using the model of the gate drive developed using the Simulink Sources and Embedded coder block set. The model build was carried out and the source code generated was compiled using CCS software. After compilation, the output file was downloaded to the memory of TI DSP TMS320F2812, which was interfaced using JTAG* to the computer in which the model was developed. The details of the TI DSP, CCS and the procedure used to generate the gate pulse, output voltage and current waveforms are explained in the following sections.

11.3.1 Digital Signal Processor

The DSP used was the one by Texas Instruments TI TMS320F2812 [6, 7]. This has four main units: (1) internal and external bus system; (2) CPU; (3) memory; (4) peripherals. The block diagram is shown in Figure 11.4a and the photograph in Figure 11.4b.

---

* JTAG is the cable used to communicate between the computer and the digital signal processor (DSP).

Interactive Model for and Real-Time Simulation 323

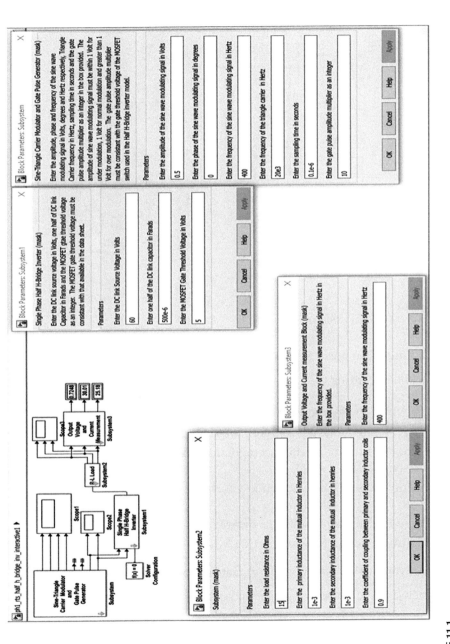

FIGURE 11.1
SPHHB PWM inverter with interactive blocks.

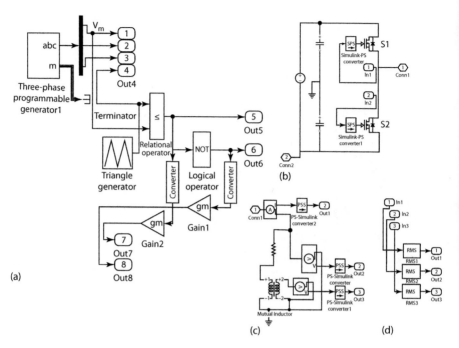

FIGURE 11.2
SPHHB PWM inverter model subsystems.

11.3.2 Code Composer Studio

The details regarding CCS are summarized as follows:

- Provides an environment for project development.
- Provides all tools required for real-time application using TI DSP.
- Programs can be written, compiled and executed either in C or in Assembler.
- Steps are Edit–Compile–Link, which are combined into Build and then Debug.
- Connects to the real-time hardware.
- Software design flow within CCS is shown in Figure 11.5.

11.3.3 Symmetric PWM Waveform Generation

Symmetric PWM waveform generation using TI DSP is given as follows:

1. Event managers:
    - The two event managers EVA and EVB provide a broad range of functions and have identical peripheral register sets.

Interactive Model for and Real-Time Simulation 325

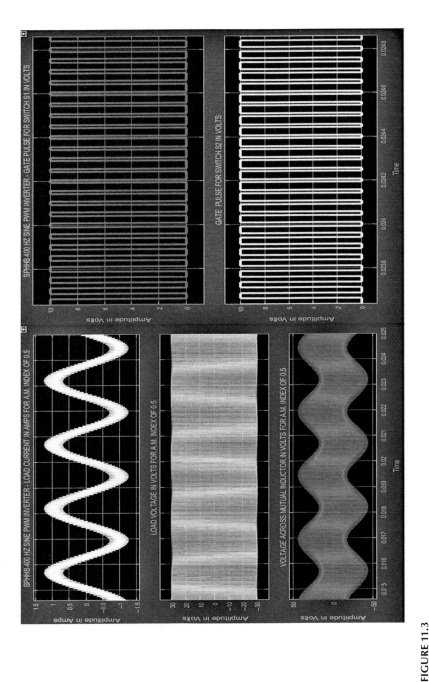

FIGURE 11.3
SPHHB sine PWM inverter: model simulation results. (a) Load current (left top). (b) Load voltage (left middle). (c) Voltage across mutual inductor (left bottom). (d) Switch S1 gate pulse (right top). (e) Switch S2 gate pulse (right bottom).

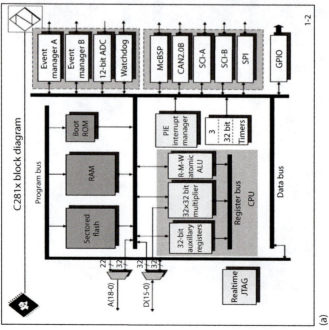

FIGURE 11.4
TI DSP TMS 320F2812. (a) Block diagram. (b) Photograph.

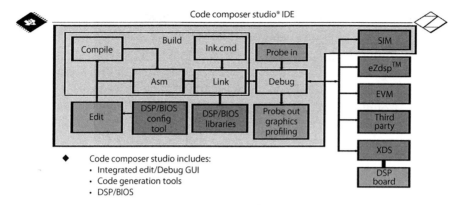

FIGURE 11.5
Code Composer Studio: software design flow.

- Event manager (EV) modules include the following:
  – General-purpose (GP) timers.
  – Full compare/PWM units.
  – Capture units.
  – Quadrature encoder pulse (QEP) circuits.
- Each EV can control three half H-bridges.
- Each EV can provide complementary pair PWM control signals.

2. GP timers:
   - Two independent GP timers are available.
   - Provide time base for the operation of compare units and for the PWM circuits to generate PWM output.
   - One readable and writable (RW) 16-bit up/down counter register TxCNT ($x = 1, 2, 3, 4$).
   - TxCNT register stores current value of the count and keeps incrementing or decrementing depending on the direction of count.
   - One RW 16-bit timer compare register TxCMPR ($x = 1, 2, 3, 4$).
   - One RW 16-bit timer period register TxPR ($x = 1, 2, 3, 4$).
   - RW individual timer control register TxCON ($x = 1, 2, 3, 4$).
   - Programmable prescaler for internal and external clock inputs.
   - Control and interrupt logic.
   - One GP timer compare output pin TxCMP ($x = 1, 2, 3, 4$).
   - Output conditioning logic.
   - GPCONA/B indicates action to be taken by timers on different timer events. Also indicate counting direction of GP timers.

3. GP timer continuous UP/DOWN counting mode:
   - Figure 11.6 shows continuous UP/DOWN counting mode.
   - Counting direction changes from UP to DOWN when (1) period register value (TxPR) is reached or (2) the value FFFF h is reached if initial timer value is greater than period register value.
   - Counting direction changes from DOWN to UP when the timer value reaches zero.
   - Timer period is 2*TxPR of the scaled clock input.
4. Symmetric PWM waveform generation:
   - In this mode (Figure 11.7), output waveform state is determined as shown as follows:
   - Zero before counting operation and unchanged until first compare match.
   - Toggles on first compare match and unchanged until second compare match.
   - Toggles on second compare match and unchanged until end of period.
   - Reset to zero at the end of period if there is *no* third compare match.

11.3.4 Sine-Triangle Carrier PWM Generation

- Here, the MATLAB/Simulink model developed using the Simulink Sources and Embedded Coder block set is used for gate pulse PWM waveform generation for the SPHHB inverter without deadband (Figure 11.8a) and with deadband (Figure 11.8b).
- Here, carrier period $Tc = 2*$(amplitude of carrier)*(CPU clock period).

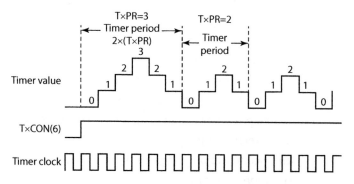

FIGURE 11.6
GP timer continuous UP/DOWN counting mode.

Interactive Model for and Real-Time Simulation

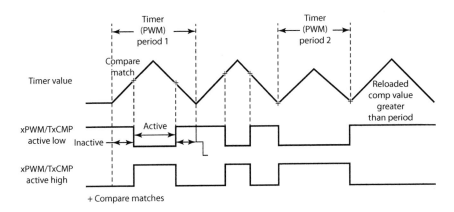

FIGURE 11.7
GP timer compare PWM output in UP/DOWN counting mode.

- For carrier frequency 20 kHz and clock frequency 150 MHz, amplitude is 3750.
- Sine wave has to be in the positive quadrant. Hence, sine-wave amplitude is 3750/2, which is 1875. Sine-wave frequency is 400 Hz.
- Sine-wave offset/bias = 1875. Data type conversion uint16 is required, as eZDSP F2812 works with unsigned integer 16 data type.
- In the C281x PWM block shown in Figure 11.8a, module A, PWM period 3750, symmetric UP/DOWN, waveform period in clock cycles and PWM1/PWM2 outputs are selected. Deadband is selected in Figure 11.8b.
- In the target preferences, SD F2812 eZdsp, TI CCS v4 and CPU clock 150 MHz are selected.
- In the Simulink model (Figure 11.8a,b), press either Ctrl-b or go to menu Tools → Code Generation → Build Model.
- Simulink RTW generates the source code of the model.
- Turn on power supply to DSP and set target configurations from 'View' menu.
- The generated code from RTW is compiled and loaded to CCS.
- After compiling source code, 'projectname.out' is generated.
- This is loaded to DSP memory by clicking 'Load Program' from Run → Load menu.
- From 'Run' menu click 'Resume'.
- The gate drive experimental set-up is shown in Figure 11.9a and the gate drive waveform is shown in Figure 11.9b for 20 kHz triangle carrier and 400 Hz sine-wave frequency with *no* deadband. The gate

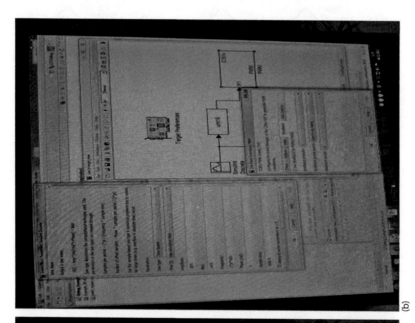

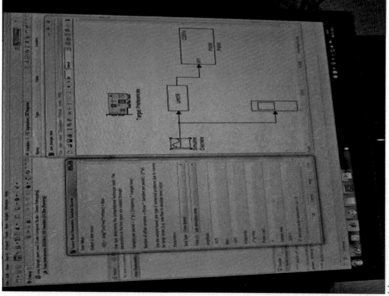

FIGURE 11.8
SPHHB sine PWM inverter gate drive model (a) without deadband and (b) with deadband.

Interactive Model for and Real-Time Simulation 331

(a)

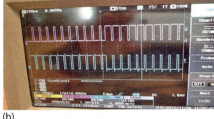
(b)

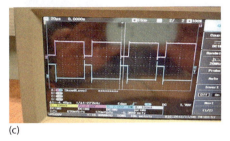
(c)

FIGURE 11.9
SPHHB sine PWM inverter gate drive experimental set-up and gate drive waveforms. (a) Experimental set-up. (b) Gate drive with no deadband. (c) Gate drive with deadband.

drive experimental result for a deadband prescale 1 and deadband period 12 is shown in Figure 11.9c [8].

- The experimental set-up of the SPHHB sine PWM 400 Hz inverter built using SiC power MOSFETs is shown in Figure 11.10 for the case with R–L Load.
- The output voltage and load current waveform of the SPHHB sine PWM inverter are shown in Figure 11.11a,b for the case with R–L load.

FIGURE 11.10
SPHHB sine PWM 400 Hz inverter experimental set-up.

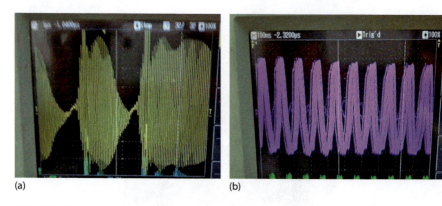

FIGURE 11.11
SPHHB sine PWM 400 Hz inverter output voltage and load current waveforms. (a) Output voltage. (b) Load current.

- The load current waveform must be divided by 15 to obtain actual load current.
- The experimental set-up of the SPHHB sine PWM 400 Hz inverter built using SiC power MOSFETs is shown in Figure 11.12a and the output voltage waveform is shown in Figure 11.12b for the case with purely resistive load.

11.4 Conclusions

Comparison of the gate drive waveform by model simulation shown in Figure 11.3d, e agree well with that shown in Figure 11.9b by real-time SIL simulation. Comparison of the load current waveform by model simulation shown in Figure 11.3a has some agreement with that by real-time SIL simulation shown in Figure 11.11b. The output voltage waveform by model simulation shown in Figure 11.3b has no perfect match with that by real-time SIL simulation shown in Figure 11.11a. One possible reason may be in the experimental set-up shown in Figure 11.10; the mutual inductor coil flux may be leaking to the surrounding atmosphere, whereas such an effect of leakage flux could not be incorporated in the interactive model shown Figure 11.1. Also, in Figure 11.3g, the load voltage waveform matches well with the one shown in Figure 11.12b by real-time SIL simulation.

Interactive Model for and Real-Time Simulation 333

FIGURE 11.12
(a) SPHHB sine PWM inverter experimental set-up for purely resistive load. (b) Output voltage.

References

1. M.H. Rashid: *Power Electronics Circuits, Devices and Applications*, Upper Saddle River, NJ: Pearson Education, Pearson Prentice Hall, 2004.
2. I. Batarseh: *Power Electronic Circuits*, Hoboken, NJ: Wiley, 2004.
3. D.W. Hart: *Introduction to Power Electronics*, Upper Saddle River, NJ: Prentice Hall, 1997.
4. N. Mohan, T.M. Undeland, and W.P. Robbins: *Power Electronics: Converters, Applications and Design*, Hoboken, NJ: Wiley, 1995.
5. The Mathworks Inc.: "MATLAB/Simulink Release Notes", R2016b, 2016.
6. Texas Instruments: "TMS320F2812 Digital Signal Processor Implementation Tutorial", 2002.
7. Texas Instruments: "TMS320x281xDSP Event Manager (EV) Reference Guide", November 2004.
8. N.P.R. Iyer: "Flexible DSP platform for power switch functional testing", Seminar on "Enabling Technologies of Power Electronics", Faculty of Engineering, The University of Nottingham, Nottingham NG7 2RD, England, 5th December 2012.

Index

AC to AC converters, 121–145, 265–272
  overview, 121
  three-phase AC controller in series
    with resistive load in delta,
    133–144, 270–272
    modelling of, 135–140
    simulation, 140–144
  three-phase AC controller with
    star-connected resistive load,
    121–132, 267–270
    modelling of, 124–129
    simulation, 129–132
AC to DC converters, 31–56, 253–256
  overview, 31
  single-phase full-wave SCR bridge
    rectifier, 38–46, 254
    with purely resistive/RLE load,
    39–44
    simulation, 44–46
  single-phase FWDBR, 31–37, 253–254
    with purely resistive/RLE load,
    32–36
    simulation, 36–37
  three-phase FWDBR, 46–56, 254–256
    with purely resistive load, 48–54
    simulation, 54–56
AC voltage controllers/regulators, 121
Amplitude modulation indices, 297, 300
AND gate, 214, 221, 227

Boost converter, 101–106, 262–265
  in continuous conduction mode,
    101–103
  in discontinuous conduction model,
    103–105
  model of, 105–106
Buck–boost converter, 106–115, 265
  in continuous conduction mode,
    106–112
  in discontinuous conduction model,
    112–114
  model of, 114–115

Buck converter, 91–101, 260–262
  in continuous conduction mode,
    91–93
  in discontinuous conduction model,
    94–95
  model of, 95–101
  in SMPS, 150–155, 272, 275, 277, 280

Cascaded multilevel inverter (CHB), 207
CCM, *see* Continuous conduction mode
  (CCM)
CCM and DCM
  boost converter model in, 105–106
  buck–boost converter model in,
    114–115
  buck converter model in, 95–101
CCS, *see* Code Composer Studio (CCS)
CHB, *see* Cascaded multilevel inverter
  (CHB)
Code Composer Studio (CCS), 321, 322,
  324
Comparators, 300
Continuous conduction mode (CCM)
  boost converter analysis in, 101–103
  buck–boost converter analysis in,
    106–112
  buck converter analysis in, 91–93
Cycloconverter, 121

DCM, *see* Discontinuous conduction
  model (DCM)
DCTLI, *see* Diode-clamped three-level
  inverter (DCTLI)
DC to AC converters, 57–89, 256–260
  overview, 57
  three-phase 120° mode inverter,
    68–80, 258–260
    line-to-line voltage analysis,
    71–73
    line-to-neutral voltage analysis,
    73–75
    model for, 76–78

335

336   *Index*

simulation, 78–80
total harmonic distortion (THD), 75
three-phase 180° mode inverter,
   57–68, 258
line-to-line voltage analysis, 58–60
line-to-neutral voltage analysis,
   60–63
model for, 63–66
simulation, 67–68
total harmonic distortion (THD), 63
three-phase sine PWM technique,
   80–84
model for, 84–86
simulation, 86–87
DC to DC converters, 91–120, 260–265
boost converter analysis, 101–106,
   262–265
in continuous conduction mode,
   101–103
in discontinuous conduction
   model, 103–105
model of, 105–106
buck-boost analysis, 106–115, 265
in continuous conduction mode,
   106–112
in discontinuous conduction
   model, 112–114
model of, 114–115
buck converter analysis, 91–101,
   260–262
in continuous conduction mode,
   91–93
in discontinuous conduction
   model, 94–95
model of, 95–101
overview, 91
Digital signal processor (DSP), 321, 322
Diode-clamped three-level inverter
   (DCTLI), three-phase, 208–214,
   290, 292
model, 209–214
simulation, 214
Discontinuous conduction model (DCM)
boost converter analysis in, 103–105
buck–boost converter analysis in,
   112–114
buck converter analysis in, 94–95
DSP, *see* Digital signal processor (DSP)

Embedded coder, 321, 322
Embedded MATLAB® function, 97, 106,
   115, 172–173, 188, 200
Event manager (EV) modules, 324, 327

Fast Fourier transform (FFT), 60, 73
FCTLI, *see* Flying-capacitor three-level
   inverter (FCTLI)
FFT, *see* Fast Fourier transform (FFT)
Flying-capacitor three-level inverter
   (FCTLI), three-phase, 214–221,
   290, 292, 294
model, 218–221
simulation, 221
Fourier coefficient, 72–73
Fourth-order converters, 91
Fourth-order DC to DC converters
   model, 165–204, 280–290
overview, 165
quadratic boost converter, 176–191,
   283
in CCM, 176–181
in DCM, 181–184
switching function concept and,
   184–191
single-ended primary inductance
   converter (SEPIC), 165–176,
   282–283
in CCM, 165–168
in DCM, 168–170
switching function concept and,
   171–176
ultra-lift Luo converter, 191–202, 283,
   290
in CCM, 191–194
in DCM, 194–197
switching function concept and,
   197–202
FWCBR, *see* Single-phase full-wave SCR
   bridge rectifier

Gate pulse generator, 18–21
General-purpose (GP) timers, 327–328

Input voltage measurement, 24–26, 27
Interactive modelling
for AC to AC converters, 121–145
overview, 121

three-phase AC controller in series with resistive load in delta, 133–144
three-phase AC controller with star-connected resistive load, 121–132
for AC to DC converters, 31–56
  overview, 31
  single-phase full-wave SCR bridge rectifier, 38–46
  single-phase FWDBR, 31–37
  three-phase FWDBR, 46–56
for DC to AC converters, 57–89
  overview, 57
  three-phase 120° mode inverter, 68–80
  three-phase 180° mode inverter, 57–68
  three-phase sine PWM technique, 80–84
for DC to DC converters, 91–120
  boost converter analysis, 101–106
  buck–boost analysis, 106–115
  buck converter analysis, 91–101
  overview, 91
for fourth-order DC to DC converters, 165–204
  overview, 165
  quadratic boost converter, 176–191
  single-ended primary inductance converter (SEPIC), 165–176
  ultra-lift Luo converter, 191–202
fundamentals, 7–29
  concept, 7–8
  overview, 7
  procedure, 8–26
of single-phase half H-bridge sine PWM inverter, 322
of SMPS, 147–163
  buck converter in, 150–155
  overview, 147
  principle of operation, 147–150
for three-phase multilevel inverters, 207–251
  CHB inverter, 221–235
  DCTLI, 208–214
  FCTLI, 214–221
  overview, 207–208

PWM methods, 238–249
RMS value and harmonic analysis, 235–238
verification, 253–319
  AC to AC converter, 265–272
  AC to DC converters, 253–256
  DC to AC converters, 256–260
  DC to DC converter, 260–265
  fourth-order DC to DC converters, 280–290
  pulse width modulation methods, 310–319
  switched mode power supply, 272–280
  three-phase five-level cascaded H-bridge inverter, 300–310
  three-phase sine PWM inverter, 297–300
  three-phase three-level inverters, 290–297
Inverters, *see* DC to AC converters

Line-to-ground voltage, 208, 214, 221
Line-to-line voltage analysis, 58–60, 71–73, 134–135, 208, 214, 221, 235–237, 300
Line-to-neutral voltage analysis, 34, 41, 60–63, 73–75, 129, 131, 300

The Mathworks Inc., 2
MATLAB® function, 243, 321
Matrix Gain, 78
Microcap11 software, 208, 214
Modelling significance and prediction, 2–3
MOSFET semiconductor switches, 303
MSLSPWM, *see* Multi-carrier sine level shift PWM (MSLSPWM)
MSPSPWM, *see* Multi-carrier sine phase-shift PWM (MSPSPWM)
MSPWM, *see* Multi-carrier sine PWM (MSPWM)
Multi-Carrier Level Shift Three-Phase Sine PWM Generator, 243–244
Multi-Carrier Phase Shift PWM Three-Phase Sine Modulating Signal Generator, 239

Multi-carrier sine level shift PWM (MSLSPWM), 238, 243–249, 310, 315–318
Multi-carrier sine phase-shift PWM (MSPSPWM), 208, 238, 239–243, 310, 313–315
Multi-carrier sine PWM (MSPWM), 208, 238, 310
Multi-Triangle Carrier PWM Generator, 243–244
Multi-Triangle Phase Shift Carrier Generator, 239

Ode23t solver, 97, 115, 129, 191, 200, 214, 221, 232, 243, 249, 300, 306, 318, 322
Op-amp model, 313–314, 317–318

Phase-to-ground voltage, 232, 237–238
Proportional–Integral (PI) controller, 147, 148, 153, 156, 163, 275, 277
PSIM simulation, 57, 70, 73, 123, 135, 148
Pulse Generator, 97, 209, 220, 244
Pulse width modulation (PWM) technique, 208
  for multilevel converters, 238–249, 310, 313–318
    multi-carrier sine level shift, 243–249, 315–318
    multi-carrier sine phase-shift, 239–243, 310, 313–315
  three-phase sine, 80–87, 297, 300
  model for, 84–86
  simulation, 86–87
PWM gate drive, 321

Quadratic boost converter, 176–191, 283
  in CCM, 176–181
  in DCM, 181–184
  switching function concept and, 184–191

Real-Time Workshop (RTW), 321
Rectifiers, see AC to DC converters
Relational Operator, 243, 313–314, 317–318
Resistor, Inductor, EMF Source (RLE) load, 21, 254

RMS, see Root mean square (RMS)
RMS2, 24, 25
RMS value and harmonic analysis, 235–238
  of DCTLI and FCTLI line-to-line voltage, 235–237
  of TPFLCHB phase-to-ground voltage, 237–238
Root mean square (RMS), 12, 21–22, 60, 67–68, 73, 78, 123–124, 250
RTW, see Real-Time Workshop (RTW)

SCHB, see Symmetrical cascade H-bridge (SCHB) inverter
SCRs, see Silicon-controlled rectifiers (SCRs)
Second-order converters, see Boost converter; Buck–boost converter; Buck converter
Selective harmonic elimination PWM (SHE-PWM), 208, 238
SEPIC, see Single-ended primary inductance converter (SEPIC)
SFWCBR, see Single-phase full-wave SCR bridge rectifier
SFWDBR, see Single-phase full-wave diode bridge rectifier (SFWDBR)
SHE-PWM, see Selective harmonic elimination PWM (SHE-PWM)
SIL, see Software in the loop (SIL) simulation
Silicon Carbide/SiC power MOSFETs, 321
Silicon-controlled rectifiers (SCRs), 121
Simulink®, 2, 7, 97, 115, 174, 191, 200, 300, 321, 322
Sine-triangle carrier PWM generation, 328–329, 331–332
Single-ended primary inductance converter (SEPIC), 165–176, 282–283
  in CCM, 165–168
  in DCM, 168–170
  switching function concept and, 171–176
Single-phase four-switch H-bridge inverters, 207

Index

Single-phase full-wave controlled bridge rectifier (SFWCBR), *see* Single-phase full-wave SCR bridge rectifier
Single-phase full-wave diode bridge rectifier (SFWDBR), 31–37, 253–254
  with purely resistive/RLE load, 32–36
  simulation, 36–37
Single-phase full-wave SCR bridge rectifier, 38–46, 254
  with purely resistive/RLE load, 39–44
  simulation, 44–46
Single-phase half H-bridge sine PWM inverter, 321–333
  interactive model, 322
  overview, 321
  real-time SIL simulation, 322–332
    code composer studio, 324
    digital signal processor, 322
    sine-triangle carrier PWM generation, 328–329, 331–332
    symmetric PWM waveform generation, 324, 327–328
Single-phase half H-bridge (SPHHB) inverter, 321
Single-phase sine-wave AC voltage, 34
Single pole double throw (SPDT) switch, 36, 44
SMPS, *see* Switched mode power supply (SMPS) modeling
Software in the loop (SIL) simulation, 321, 322–332
  code composer studio, 324
  digital signal processor, 322
  sine-triangle carrier PWM generation, 328–329, 331–332
  symmetric PWM waveform generation, 324, 327–328
Space vector PWM (SVPWM), 208, 238
SPDT, *see* Single pole double throw (SPDT) switch
SPHHB, *see* Single-phase half H-bridge (SPHHB) inverter
SVPWM, *see* Space vector PWM (SVPWM)

Switched mode power supply (SMPS) modeling, 147–163, 272, 275, 277, 280
  buck converter in, 150–155
  overview, 147
  principle of operation, 147–150
Switching function concept, 1
  and quadratic boost converter, 184–191
  and SEPIC, 171–176
  and ultra-lift Luo converter, 197–202
Symmetrical cascade H-bridge (SCHB) inverter, 207
Symmetric PWM waveform, 324, 327–328

Texas Instruments (TI), 321, 322
TFWDBR, *see* Three-phase full-wave diode bridge rectifier (TFWDBR)
THD, *see* Total harmonic distortion (THD)
Three-phase 120° mode inverter, 68–80, 258–260
  line-to-line voltage analysis, 71–73
  line-to-neutral voltage analysis, 73–75
  model for, 76–78
  simulation, 78–80
  total harmonic distortion (THD), 75
Three-phase 180° mode inverter, 57–68, 258
  line-to-line voltage analysis, 58–60
  line-to-neutral voltage analysis, 60–63
  model for, 63–66
  simulation, 67–68
  total harmonic distortion (THD), 63
Three-phase AC controller in series with resistive load in delta, 133–144, 270–272
  modelling of, 135–140
  simulation, 140–144
Three-phase AC controller with star-connected resistive load, 121–132, 267–270
  modelling of, 124–129
  simulation, 129–132

340   *Index*

Three-phase AC voltage source, 9, 12–16
Three-phase five-level cascaded
    H-bridge inverter (TPFLCHBI),
    221–235, 300, 303, 306
  model, 225–232
  simulation, 232–235
Three-phase full-wave diode bridge
    rectifier (TFWDBR), 46–56,
    254–256
  with purely resistive load, 48–54
  simulation, 54–56
Three-phase half-wave SCR converter, 9,
    16–18, 27
Three-phase MOSFET inverter
  model, 300
Three-phase multilevel inverters,
    207–251
  CHB inverter, 221–235
    model, 225–232
    simulation, 232–235
  DCTLI, 208–214, 290, 292
    model, 209–214
    simulation, 214
  FCTLI, 214–221, 290, 292, 294
    model, 218–221
    simulation, 221
  overview, 207–208
  PWM methods, 238–249, 310, 313–318

multi-carrier sine level shift,
    243–249, 315–318
multi-carrier sine phase-shift,
    239–243, 310, 313–315
RMS value and harmonic analysis,
    235–238
  of line-to-line voltage, 235–237
  of phase-to-ground voltage, 237–238
Three-phase sine PWM inverter, 256
TI, *see* Texas Instruments (TI)
TMS320F2812, 321, 322
Total harmonic distortion (THD), 63,
    75, 221
TPFLCHBI, *see* Three-phase five-level
    cascaded H-bridge inverter
    (TPFLCHBI)
Transport Delay, 244, 249
Triangle carrier pulse, 150, 153, 275, 315

UCHB, *see* Unsymmetrical cascaded
    H-bridge (UCHB) inverter
Ultra-lift Luo converter, 191–202, 283, 290
  in CCM, 191–194
  in DCM, 194–197
  switching function concept and,
    197–202
Unsymmetrical cascaded H-bridge
    (UCHB) inverter, 207

PGSTL 03/16/2018